D0458956

Agriculture: Foundations, Principles and Development

Agriculture: Foundations, Principles and Development

J. R. RAEBURN

FRSE, FIBiol, PhD(Cornell)

A Wiley–Interscience Publication

JOHN WILEY & SONS LTD
Chichester · New York · Brisbane · Toronto · Singapore

Library of Congress Cataloging in Publication Data:
Raeburn, J. R.
 Agriculture—foundations, principles and development.

 'Wiley–Interscience publication.'
 Includes bibliographies and index.
 1. Agriculture. I. Title.
S495.R24 1984 630 84-3619

ISBN 0 471 10308 X

British-Library Cataloguing in Publication Data:
Raeburn, J. R.
 Agriculture: foundations, principles and development.
 1. Agriculture
 630 S493

ISBN 0 471 10308 X

Typeset by C. R. Barber & Partners (Highlands) Ltd., Fort
William, Scotland.
Printed by Pitman Press Ltd., Bath, Avon.

To all who have striven and all who strive,
on farms or elsewhere, for the improvement
of agriculture

Contents

PART III   THE WORK SUB-SYSTEM

PART IV   THE FARM-ECONOMIC SUB-SYSTEM

PART VI   INTRODUCTION TO POLICIES

Preface

In most countries, modern trends and problems are raising the value of the understanding of agriculture as a whole. Farmers rely increasingly on purchases of chemicals, machinery, well-bred plant material, and other goods and services. The *knowledge* that guides these purchases comes not only from farmers' interpretations of their own experiences, but also from other farmers, and from many different specialists in natural sciences, engineering, farm economics, and other subjects. Farmers have to make decisions in their own circumstances, using many different kinds of knowledge. And because farmers rely on *markets* for sales of outputs as well as for purchases of inputs, this knowledge includes price and other changes in markets. Almost all *governments* affect directly or indirectly both the build-up of knowledge and the conditions in markets. This is because governments are concerned with a wide range of issues related to agriculture, including human nutrition and security of food supplies, food quality and hygiene, population increases, economic growth, inflation, marketing costs, international trade, balancing international payments, employment, rural incomes, rural–urban migrations, land tenure, land-use planning, conservation of natural resources, recreation in rural areas, animal welfare, and pollution control. All these issues raise many questions that require understanding of several different aspects of agriculture, of their interrelationships, and of variations in agriculture from place to place, and changes over time. Naturally in all this complex the attentions of politicians, 'government circles', social scientists, and 'thinking publics' tend to be concentrated, at particular times and places, on narrow questions. But the answers can seldom be satisfactory unless there is a wider understanding.

Naturally also, as the need for understanding increases, the difficulties in meeting it increase. The rapid expansions of *knowledge* by specialists in particular subjects and by farmers in particular areas, the increasing importance of *market* changes in the world as a whole, and the wide ranges over which *governments'* priorities can vary – all these make more difficult the provision of an overall view that will be useful in different circumstances, and continue to be useful. Even in college and university courses in agriculture, shortage of time has resulted in much specialization. Opportunities are now strictly limited for adequate introductions to the whole complex of agriculture, and adequate revisions and reassessments of it after particular specializations and before courses end. The difficulties are all the greater where many students have little or no experience of life and work on farms. Courses in the natural sciences, geography, economics, statistics, sociology, law, politics, government, and other subjects can devote next to no time specifically to the understanding of agriculture as a whole, even although many graduates from these courses take up responsibilities that cannot well be carried out without such understanding – e.g. in agricultural research, administration in central and local governments, and management in much of commerce and industry. Once they have these responsibilities their opportunities to secure an overall understanding are increased in some respects, but limited in others because they have little time and, commonly, inadequate foundations in biology, or farm operations, or farm economics on which to build. They therefore encounter serious obstacles in trying to piece together useful basic understandings from the wide range of specialist books, from books descriptive of the farming systems of particular times and places, or from encyclopaedias. Even more difficult is the position of many politicians and general readers who have even less time available and whose previous studies contributed little to the understanding of agriculture. Most farmers too have difficulties. To be able to assess well their own experiences with crops and animals, the flows of new biological knowledge and equipment designs, the changes in market prices and conditions, and the changes in government policies – all this requires as early as possible in farmers' careers far better basic understandings than can readily be secured.

The purpose of this book is therefore to provide an overall picture of agriculture that is useful as an introduction, and as a general guide and frame of reference when particular questions or further studies are pursued, when information is stored, and when revisions and reassessments are desirable.

The method used depends first on a brief review in Chapter 1 of nine questions that are central to the world's problems about agriculture. Then the whole complex is divided into three sub-systems – biological, work and equipment, and farm-economic – and the relations of these to the general socioeconomic system. The method also depends on the sequence of presentation; basic principles rather than descriptive detail, but with telling examples; many cross references in the text; a few questions and exercises at the ends of chapters that require the relation of information from readers' own areas or countries to the whole concept; and finally a balanced selection of further readings. The emphasis is on interconnections within and between the sub-systems; the importance of markets and other components of socioeconomic systems; variations in many biological aspects between sites, seasonally, and from year to year; differences in equipment and physical work methods; and the importance of human attitudes, skills, information, and decisions. *Developments* are determined by all of these.

The space allotted to different chapters is partly determined by whether the respective principles and illustrations are available to most readers. Thus, for example, relatively little space is allotted to the basis of heredity in Chapter 4, but more to the attributes to be improved in crop plants. Yet the essence of all principles of great importance for decisions is set out, so that all readers may readily have them.

Considerable space is used for examples from widely contrasting biological, economic, and social circumstances. The purposes in this are to help towards sufficient worldwide understandings, and to reaffirm an old saying – that if you study the agriculture of other countries, you understand better that at home.

Another guide in allotting space is the importance of liaison and team work, especially that between subject specialists and farmers. A base is required of mutual interest and understanding that is wider than any specific questions in mind.

Readers who want a quick introduction may find it convenient to concentrate first on Chapters 1 and 2 and the first chapters of the later parts – that is, Chapters 3, 7, 10, 13, and 21. Then, to pursue particular questions, they can then use the *list of contents* and the *index*, as well as cross references, and suggested readings.

The further readings suggested at the ends of chapters include some major works (marked †), some that are suited to the later years in college and university courses (marked *), and some that may be so used but are also suited to the general reader (not asterisked). Useful bibliographies are included in these readings, and the continuing abstract services of the Commonwealth Agricultural Bureaux are also listed here.

Any book of this kind must draw on a great body of knowledge, built up by the endeavours of thousands of men and women over many years and in many countries. Hence the dedication, because no other acknowledgement is possible here. But special thanks are due *for encouragement* to former students and colleagues and the publishers and their consultants, for *financial help* in the form of an Emeritus Fellowship in the University of Aberdeen to the Leverhulme Trust, *for library facilities* to the School of Agriculture, Aberdeen, and the Institute for Agricultural Economics, Oxford, *for research assistance* to C. H. Liau, and *for helpful criticisms* of particular chapters to the following: Sir David Cuthbertson, G. E. Dalton, D. Macdonald, A. D. McKelvie, R. Passmore, R. T. Pringle, J. K. Thompson and J. H. Topps. I am also very grateful for *guidance and data for example areas* to C. Barlow, C. A. Bratton, H. C. Halcrow, R. C. Huffman, R. J. Isaacs, R. W. Herdt, C. S. Ng, J. G. Ryan, P. J. Thair, I. Sturgess, and their colleagues. Any remaining errors are of course the author's responsibility. Special thanks are also due *for typing* to Jean Revolta, Margaret Sinclair, and Barbara Rae.

Aberdeen, 15 January 1983   J.R.R.

Part I

Introduction

Chapter 1

The importance, trends, and problems of agriculture

At the core of the many detailed problems of agriculture is a set of nine general questions. A brief review of these helps to show the importance of agriculture, the many-sided nature of past developments, and why many types of change are required for the future.

The first question is about the kinds of contribution that agriculture makes to human welfare. Three questions concern the demand for foodstuffs; three concern their supply. The eighth question is about the industrial raw materials that agriculture produces. And the ninth is about the place of agriculture in social organization and progress.

1.1 WHAT DOES AGRICULTURE CONTRIBUTE TO HUMAN WELFARE?

1.1.1 Outputs

The main contribution is food. Mankind depends for almost all supplies on agriculture. Hunting and fishing make comparatively small contributions, and that from artificial synthesis is very small. Other outputs are raw materials for beverages – coffee, tea, cocoa, grain for malting, grapes for wine making, and others. Some raw materials for industry are now synthesized artificially and compete with agricultural products such as cotton, other vegetable fibres, wool, silk, hides, rubber. But agricultural production of raw materials is still important. Exports are worth more than US$18 000 million a year. There are also timber and fuel from woodlands and forests, and fish from fish farms.

1.1.2 Employment

In agrarian societies with little modern industry, the majority of families has agriculture as the main employment. Other activities take up much time but are so interwoven with agriculture that the net employment in agriculture is seldom well measured. The statistics available are the percentages of the economically

3

active that have agriculture as their main source of income (Table 1.1, column 4). After productivity in agriculture rises, non-agricultural activities expand, and labour becomes more specialized, agriculture and other economic sectors become more clearly interdependent. Inputs to farms increase – mechanical equipment and power, chemical fertilizers and pesticides, services in processing and distribution of feedingstuffs and seeds, and others. Farm products have to be marketed, and their processing and packaging are done more and more off farms. Thus in the 'middle income' and 'industrialized' countries much off-farm employment is in activities that are closely related to agriculture, but outside the statistical definitions of it. So percentages of the economically active populations recorded as in agriculture tend to underestimate the total employment related to it.

Table 1.1   Contributions of agriculture to total gross domestic products, employment, and exports[a].

1	2	3	4	5	6	7
	GDP, 1979		Labour force in Agriculture		Exports, 1979	
Types of economy	Total per capita	From agri-culture	1979	Reduction from 1960 to 1979	Agri-cultural	Share of total exports
	US$	per cent of total	per cent of total		US$'000 million	per cent
Low income	230	38	71	5	} 60	20
Middle income	1 420	14	43	15		
Industrialized, market	30 430	4	6	10	128	12
Capital surplus oil exporters	5 470	2	44	14	+	+
Centrally planned Asian	260	31	71	n.a.	4	27
Other	4 230	15	17	24	10	8
World	...	...	46	...	202	13

Sources: Based on World Bank (1981) and FAO (1982).
[a]Including contributions from forestry, fishing, and hunting.

1.1.3   Structural changes

The low percentages in industrialized economies do, however, indicate how great were past changes in the structure of production. Without expansion in the outputs of foodstuffs and raw materials, and increases in outputs per unit of labour in agriculture, so much labour could not have shifted to other activities, nor specialization developed so far. During the early stages of industrialization much of the capital required was obtained from agriculture. The increases in overall production that industrialization and specialization made possible were immense. The shifts of labour out of agriculture were substantial even in industrialized economies during the 1960s and 1970s. (Table 1.1, column 5). And agriculture's increasing dependence on other economic sectors expanded the demands for their products.

We should also recognize that structural changes have included some big changes in *where* agricultural production takes place (e.g. grain, fruit, and meat production). Such alterations have, like other structural changes, contributed to the development of markets for production goods and services and to overall productivity.

1.1.4 Foreign exchange

The benefits that can result from structural changes cannot be secured in full without trading between nations. Trading allows freedom to relate the pattern of agricultural production not simply to the locations of consumers but much more closely to the pattern of advantages in production (as determined by climates, soils, technological skills, and equipment), and to the pattern of marketing costs. Production can therefore be more efficient. Trading also permits worldwide improvements in efficiency through economic use of machinery, production materials, and ideas that would otherwise be restricted geographically. Developments of the use of fertilizers and pesticides, tractors, and modern processing machinery have all required international trade. Moreover, trading often makes supplies of foodstuffs and raw materials more secure (e.g. when supplies can be drawn from other countries because the harvest in a particular country is poor).

The earning of foreign currency by exports is of course necessary to permit purchase of imports. And in many poor countries the need for this is all the greater because economic growth has to include many changes in transport, education, mining, manufacturing industry, and other sectors, all of which require some foreign currency. In most low and middle income economies, and in some centrally planned ones, agriculture has to contribute much to the total earnings because much of the national product is agricultural. In some industrial economies such as the USA and Canada the foreign exchange earnings of agriculture are still important to the whole economic structure, as well as to particular rural communities. And in other industrialized economies that are even more dependent on international trade than the USA and Canada special efforts are made to try to secure balances between agricultural and other activities as affecting foreign exchange requirements and earnings. An indication of the value of crop and animal product exports is given in Table 1.1, columns 6 and 7.

1.1.5 Net returns

As already indicated, the definitions of agriculture used in governments' statistics and national accounting exclude many activities closely related to production on farms. And where a title such as 'food and agriculture' is used, the boundaries with manufacturing and other activities may still not be satisfactory. This is one major difficulty in finding useful measures of the net returns to agriculture. A second difficulty is that the greater part of labour on farms is by farmers and their families, and it is they who provide also the land, capital, and management. The net returns to these various components often cannot be well separated (see Section 12.1.3). Third is the difficulty of valuing (i) produce that is consumed on the farms where grown, and (ii) the use of farmhouses, vehicles, etc. A quick general indication of returns can be obtained from the calculated contributions of agriculture to gross domestic products, which are aggregate measures of total national outputs of goods and services (Table 1.1, column 3). These contributions are equal to the returns for all labour, the use of land and capital, and all elements in management. They still include amounts that should be used to offset depreciation of machinery, buildings, etc. The returns as calculated are large parts of the total gross domestic products in India and other low and middle income countries and in the USSR (17 per cent). When all the definitions are borne in mind, the figures for industrialized market economies are also substantial.

The returns expressed as a percentage of the gross domestic product are often compared with the percentage of the total labour force that is in agriculture (Table 1.1, column 4). In low and middle income countries, and in the main oil exporters, the returns seem very low. They also seem low in some industrial countries. But tentative conclusions require closer study (see Sections 16.2 and 16.3).

1.1.6 Economic stability

Agriculture contributes to the security of food supplies through increases in total production and by trading, both in and between nations. The biological nature of agricultural production opposes short term shut-downs (Section 14.4.2). Also in most economies, security of energy supplies in human foodstuffs is obtained by producing a variety of crops and by making possible, when there tend to be shortages, a switch of feedingstuffs from animals, and additional slaughter of animals.

The dispersed nature of agriculture fosters the provision of social security by families for the rural unemployed, landless, and underemployed.

But the limits to the security provided should be recognized. The instabilities in agriculture can be serious. And neither agrarian societies in low income countries nor complex agricultural businesses in industrialized countries can, without serious disruption, provide enough security, tax revenues, and charity to meet all possible needs.

1.1.7 Social benefits

The above types of contribution are commonly called 'economic' but they all have major importance for social relationships within and between nations, and for social progress. In addition there are social benefits that are difficult to measure in economic terms. They arise from the satisfaction by agricultural activities and rural life of basic human instincts and desires that cannot be so well satisfied otherwise. Even after centuries of urbanization and contacts between races and nations, the evolution of human cultures has sustained these instincts and desires in large groups in all countries. Commonly the slow pace of exits from agriculture, despite comparatively low incomes, indicates substantial non-material benefits from agriculture. Other evidence comes from (a) much poetry and other literature; (b) the maintenance of many part-time farms in industrialized economies; (c) the high prices paid for farms in Europe and North America by men successful in other industries; (d) the large sums that taxpayers will pay to avoid more rapid rural depopulation, or to preserve landscapes previously largely created by agriculture.

1.2   HOW MANY PEOPLE WILL BE FED?

In 1980 the total human population was 4400 million, and the yearly increase was 76 millions (1.7 per cent). Increases in recent decades have been as follows:

1920s	204 million	1950s	514 million
1930s	225 million	1960s	651 million
1940s	222 million	1970s	754 million

These increases had been reduced by under-nutrition and wars in much of the world and by severe famines in particular countries. But even so, such accelerating increases were much larger than those ever before experienced in human history.

What has been called the 'population explosion' continues. The total human population is expected to increase by another 1642 million (37 per cent) during the 20 years from 1980 to 2000 (Table 1.2). If recent trends in birth and death rates continue, the total will then increase by a further 3738 million (62 per cent) before becoming stationary (Table 1.2).

Some clues to the causes of the 'explosion' are provided in the statistics of birth and death rates in different groups of countries (Table 1.2). The low income group had, in 1978, crude annual birth rates averaging 39 per 1000 of population, but crude death rates of 15. The difference of 24 provided a quick indication of the pace of population increase. Death rates had been reduced by much more than birth rates. The difference had indeed become greater than any experienced by Europe or Japan during their 'population explosions' of the nineteenth century. Death control through control of malaria and other diseases of the Tropics and Sub-tropics had precedence over birth control. In the middle income countries the differences averaged in 1978 also 24; in the oil-exporting countries with capital surpluses, 29; in China, 12; in the USSR, 8; in the industrialized market economies, only 5.

Calculations of future population increases depend of course on assumptions about future changes in birth and death rates. Such changes depend on (a) past rates, because these have determined (i) the numbers of people already born into

Table 1.2   Birth and death rates and estimated population increases in the world:[a] 1980, 2000, and beyond.

1	2	3	4	5	6	7	8
		Crude annual rates[c]			Increases in population		
Types of economy	Population in 1980[b]	Birth	Death	Net increase	1980	to 2000	2000 to first year stationary
	millions	*number per 1000 population*			*millions*	*per cent*	*per cent*
Low income	1348	39	15	24	702	52	99
Middle income	916	35	11	24	493	54	84
Industrialized	673	14	9	5	63	9	5
Capital surplus							
oil exporters	64	43	14	29	40	63	95
Centrally planned							
China	977	18	6	12	262	27	26
USSR	267	18	9	8	47	18	13
World total[a]	4387	...	...	...	1642	37	62

Sources: World Bank (1980, 1981).
[a]Not including countries with less than 1 million population.
[b]Partly forecast.
[c]Estimated for 1979.

and surviving in different age groups, and therefore (ii) the number of women of child bearing ages; (b) future personal decisions about limiting numbers of children in families; (c) future effects on death rates of public and other measures for health, of wars and other disasters, and of food shortages. Therefore the assumptions on which future populations are calculated should include judgements of potential changes in family sizes and of various determinants of death rates.

Some indications of the range of possibilities are set out in Table 1.3. In the low and middle income countries and in the oil exporters, high birth rates were the result of decisions to have an average of five, six, or more children per family (Table 1.3, column 2). But such decisions were changing slowly (Table 1.3, column 4). In China the reduction was rapid (18 per 1000 population to reach a birth rate of 18 by 1978). On the other hand, deaths were reduced more than births in the oil exporters, and equally with births in the low income group (Table 1.3, columns 4 and 8). The scope for further reductions in death rates was substantial (Table 1.3, columns 5–8).

Demographers emphasize, therefore, that calculations of future populations are no more reliable than the assumptions on which they are based. When populations may become stationary and at what levels are therefore speculative (Table 1.2, column 8). And even the predictions of the total human population increase by AD 2000 range widely around the 1642 million in Table 1.2.

Table 1.3   Human fertility and mortality in the world:[a] 1978 and reductions from 1960 to 1978.

1	2	3	4	5	6	7	8
Types of economy	Total fertility rate 1978[b]	Crude annual birth rate		Mortality rates of those aged 1–4 years		Crude annual death rate	
		1978	Reduction	1978	Reduction	1978	Reduction
	number	*number per 1000 population*		*number[c]*		*number per 1000 population*	
Low income	5.4	39	9	20	10	15	9
Middle income	4.9	35	5	10	8	11	3
Industralized	1.8	14	6	1	0	9	1
Capital surplus oil exporters	6.5	43	5	16	13	14	7
Centrally planned							
China	2.3	18	18	1	13	6	5
USSR	2.4	18	6	1	0	9	2[d]

Source: As for Table 1.1.
[a]As for Table 1.2.
[b]Number of births per woman if she lived to the end of her child-bearing years.
[c]Per 1000 in the same age group.
[d]Increase.

But on all reasonable assumptions the numbers to be fed in AD 2000 will be far greater than in the 1970s. And beyond, yet larger numbers will arise.

1.3   FAMINE AND UNDER-NUTRITION: WHAT ARE THE EXTENTS?

Severe famines have been suffered often in human history. The worst this century were in China (1916 and 1920); the USSR (1905, 1921, and 1932–3); Bengal (1943); Biafra, Nigeria (1969–70); Bangladesh (1971–3); Africa, from the south Sudan to Senegal, the Sahel (1968–74); and Kampucea (1979). To avoid serious famine during 1965–7 in India and Pakistan some 60 million tons of special grain imports were required. The Sahel famine was not so promptly countered. But now the Food and Agriculture Organization's early warning system reports countries that are in imminent danger of serious food shortages. At any one time, a dozen or more countries are reported as threatened.

Under-nutrition that is more continuous than famine is widespread in the poorer countries. Medical studies show that several million new clinical cases of protein–calorie malnutrition (PCM) result each year. Where there is deficiency of calories (and related protein, minerals, and vitamins) growth is retarded and muscle and fat waste away. Gastroenteritis, respiratory infections, measles, tuberculosis, and severe parasite attacks, and other chronic diseases, are frequent. The disease is called *marasmus* and most commonly affects infants at about 8 months old. Where the deficiency is of protein and related minerals and vitamins but not of calories, *kwashiorkor* results, commonly at 18 months old, with similar effects on growth and muscles, and miserable apathy. A considerable proportion of pre-school children in the poorer countries have sub-clinical forms of PCM.

From a medical standpoint, human reproduction is, even under nutritionally quite adverse conditions, efficient enough to increase populations rapidly. The reductions in death rates made so far have therefore brought us to an 'era of surviving children'. 'Large numbers survive but, because of under-nutrition of mothers and young, are left vulnerable to subsequent infection and disease, and are unable to achieve optimum benefit from schooling' (Jelliffe, 1973).

Medical studies and detailed surveys of food consumption show that under-nutrition is commonest (a) where climates and soils do not favour agricultural production, and transport is not well developed; (b) among unemployed and casual labourers – landless people, and those with very little land; (c) among the nutritionally vulnerable groups, especially pregnant and lactating women, infants and young children. Also, for physical work, additional calories are required, but are not fully available. Group feeding experiences in armies, railway and other construction projects, and plantations confirm the great practical importance of conclusions from physiological studies. These are, in brief, that 'in countries with an insufficient amount of food available, it cannot be expected that efficient manual work can be performed' (Åstrand, 1973).

In the 'developed' economies, under-nutrition still occurs among some of the very poor, and among those with particular metabolic diseases. The problems arising are not now for agriculture so much as for medicine and social administration.

Because of the importance of plans to eradicate under-nutrition in the 'developing' countries, but the high costs of medical surveys, statistical attempts are made to measure the extents of under-nutrition and monitor progress. Valid comparisons of nutrient requirements and available nutrient supplies are, however, difficult to secure.

Nutrient requirements vary with air temperatures, age, sex, past physical growth and development, pregnancy and lactation, physical activities (including agricultural work), disease, and other variables. Between individuals who are otherwise apparently similar, there are variations in basic metabolic rates, in reactions to changes in energy ingested (consumed), and in physiological absorption. Because under-nutrition stunts body growth and reduces mental and physical activities, it tends to reduce future use of nutrients. There can therefore be widely different opinions about what standards of 'requirements' should be, as well as great difficulties in securing measures of the basic determinants.

Nutrient supplies should be measured as those actually consumed. But statistics of foodstuff production, storage, processing, and marketing – and those of food purchases and 'losses' in households – are seldom precise enough.

Comparing supplies and requirements may be misleading unless within nations the pattern of consumption rates can be contrasted with the pattern of requirements – that is the dispersion of population in sufficiently detailed requirements categories. Average calorie supplies and requirements per caput for a whole nation may be poor indicators of the occurrences of under-nutrition.

The FAO/WHO standards for national average daily calorie intake requirements take account of differences in average air temperatures, age and sex distributions, and body weights, and assume moderate physical activities. Some observers have judged these standards too high for many countries, because based on samples too limited in scope as to numbers, races, and levels of health. The calorie supply averages in different countries may also be subject to error. Even so, the national shortages in the countries most seriously affected by poverty – particularly India, other Far Eastern countries, and Africa – seem so big that, supported by the medical evidence, they call for study and action (Table 1.4, columns 2–4).

Table 1.4   Average calorie supplies, and estimates of under-nourished populations in FAO regions.

1	2	3	4	5	6	7
		Calorie supplies, 1972–4		Estimated populations with energy intakes		
Region	Total population in 1978	Average per caput per day	Percentage of average requirements	Below IFRPI standard	Below FAO critical limit, 1972–4	
	millions	*calories*	*per cent*	*millions*	*millions*	*per cent*[b]
Developing market economies						
Africa	348	2110	91	180	83	28
Latin America	347	2540	107	70–120	46	15
Near East	201	2440	100	70–100	20	16
Far East	1183	2040	92	590–600	297	29
Total[a]	2084	2180	95	910–1000	455	25
Centrally planned economies						
Asia	954	2290	97	n.a.	n.a.	n.a.

[a]Including in columns 2–4 data for countries with 5 million population not included in the regions named.
[b]Of total populations in 1972–4.

Other statistical calculations have lower standards of requirements, but allow for income distribution patterns. The International Food Policy Research Institute (IFPRI) estimated, using World Bank calculations within nations, the number of people who had in 1975 daily food energy intakes 200 or more calories below the FAO/WHO standards (Table 1.4, column 5). The FAO in 1977 set still lower standards. These were 1.5 times basic metabolic rates and so allowed, on average, no more than the energy costs of body maintenance. Nothing was allowed in the averages for other activities. Moreover, to allow for some individuals' basic metabolic rates varying downwards from the averages, the standards were reduced by 20 per cent to secure the 'reasonable critical intake limits'. Nation by nation, more detailed income and food consumption patterns were then used to calculate how many people had less than their 'critical limits' (Table 1.4, columns 6 and 7). In developing market economies as a whole, the energy in the food of 25 per cent of the total population – that of some 455 million people – appeared to be below their 'critical limits'. The IFPRI calculations indicated that the total population 'under-fed' was between 910 and 1000 million. Both sets of calculations showed that the shortages were greatest in Africa and the Far East (including India, but with no data for China).

Such results make highly desirable the improvement of all the types of data used and of the methods of statistical analysis. But, with the medical evidence, they confirm that great increases in the foodstuffs available in the developing countries for undernourished groups are necessary, if anything like full human development and health are to be achieved.

If the increases in supplies measured in calories were achieved, the shortages in protein, minerals, and vitamins would probably also be almost entirely overcome. The typical cereal diets of the poor would ensure this. But medical-nutritional evidence shows that special care would still be required about (i) the protein, mineral, and vitamin intakes of pregnant and lactating women, and children on weaning from breast feeding; (ii) diets in which roots and tubers provide much of the energy but little protein; (iii) vitamin A deficiency causing poor sight and blindness; (iv) iodine deficiency causing goitre.

1.4 FOOD CONSUMPTION PATTERNS AND OVER-NUTRITION: WHAT CHANGES SEEM PROBABLE IN EFFECTIVE DEMANDS FOR FOODSTUFFS?

This question turns our attention from nutrient requirements based on human physiology to effective demands, based on preferences and abilities to buy (or produce for self-supply).

Our preferences are of course governed to some extent by our physiology. Our inborn appetite conrols seem to be mainly those that helped our prehistoric ancestors to survive as hunters and gatherers, with sometimes little and sometimes much to eat. Our appetites naturally lead us to eat more than we need. If our diets are made appetizing through variety, textures, flavour, cooking, and presentation, energy intakes are raised further, and protein intakes much further, than are required nutritionally. The FAO averages for developed market economies indicate that the energy in food supplies bought by consumers totals about 20 per cent more than their requirements.

But there are physiological limits to our appetites, and in determining our actual diets they operate along with our detailed preferences and purchasing powers. This can be seen from the relations of food consumption to income differences and changes. Such relations are of fundamental importance to

Table 1.5   Income elasticities of demand [a].

	India	Nigeria	Japan	UK	Canada	USA
	per cent increase in demand per 1.0 per cent increase in income					
All cereals	0.4	0.4	0.0	−0.5	−0.4	−0.2
Pulses, nuts	0.5	0.3	0.1	0.0	−0.1	−0.1
Vegetables	0.7	0.6	0.3	0.3	0.1	0.2
Fats, oils	1.0	0.6	0.9	0.0	−0.1	0.0
Eggs	1.5	1.0	0.8	0.3	−0.1	−0.1
Milk[b]	1.6	1.0	0.8	0.0	−0.2	−0.2
Meat	0.5	0.8	1.1	0.4	0.3	0.2

Source: FAO (1967).
[a]For quantities of foodstuffs.
[b]Including milk products, excluding butter.

agriculture. The demand for food does expand as incomes rise but, when more and more food is taken, additional food has less and less value relative to other goods and services. Thus in 1977 in the UK, household expenditures for food were higher by only 0.14 per cent with every 1.00 per cent by which income per head was higher in one income group than in another. Such percentage figures are called income elasticities of demand. Table 1.5 compares estimates for different kinds of food in countries with different levels of income per head (Table 1.6).

Agriculture is much affected because when incomes rise consumers alter the compositions of their diets. Foods with higher income elasticities of demand (e.g. meats) are preferred more than those with lower (e.g. bread). And similarly the better qualities of each food are preferred (e.g. of cheese or fruit). Consumers also alter preferences about the services added to their foodstuffs. At one extreme, for

Table 1.6   Relation of food supply to national incomes and other national differences.

	Gross national product, 1978	Food supply, 1977–9		
		Energy	Protein	
			Crop	Animal
	US$ per head	*calories per head per day*	*grams per head per day*	
India	180	1996	44	4
Nigeria	560	2295	44	6
United Kingdom	5030	3275	36	50
Japan	7280	2847	45	42
Canada	9180	3346	36	66
USA	9590	3537	34	72

Sources: World Bank (1980) and FAO (1981, 1982).

example, live fowls are bought by poorer families direct from farmers, and killed, prepared, and cooked at home. At the other extreme, selected chicken joints are served, well cooked and garnished, to the customers of city restaurants. The income elasticities of demand for *quantities* at points of first sale by farmers can differ widely from those for *values* of foods at retail or in restaurants.

Social habits and experiences, and changing food technology, also determine food consumption patterns. And preferences can alter over time due to education, advice, advertising, urbanization, contacts between cultures, and changes in employment (including female employment). Therefore in explaining past and forecasting future effective demands for foodstuffs we have to take account of more than population and income changes.

Table 1.6 sets out some international differences in calorie and protein consumption and in types of food as measured by the ratio of animal to plant sources of the protein. Higher calorie consumption was related to higher incomes, except that Japan had relatively low consumption because of food habits, human body weights, and other reasons. Japan also had relatively low protein consumption. But we should note that with higher incomes it was the animal sources that became more important and Japan's income elasticities of demand for milk, meat, and eggs were high (Table 1.5).

Future effective demands in low and medium income countries will be influenced by continuing urbanization, advertising, and other culture contacts. More white bread and rice will be preferred to sorghum, barley, and hand-milled rice. The use of processed baby-foods may increase. Such changes can have serious nutritional effects. But in practice the effects on agriculture will be small as compared with the great expansion in demand due to 'population explosions' and income increases. In 1965–8 the FAO calculated that effective demands in the developing countries (excluding China) might rise during the 23 years between 1962 and 1985 by 82 per cent due to population increases, and by 33 per cent because of growth of incomes per caput – 142 per cent in all $[(182 \times 1.33) - 100]$. This would be equivalent to increases of 3.9 per cent a year: cereals, 3.1 per cent; starchy roots, 3.2 per cent; vegetables, 4.5 per cent; meat, poultry, fish, 5.6 per cent. The growth of incomes has been less than calculated, but any valid changes for this reason (or because of unrelated changes in food preferences) would still leave such rates of demand increase very high.

The future of effective demands in industrialized economies will be different. Population increases and income elasticities of demand for foods as a group are low. And consumers' preferences may be altering because of medical advice against high consumption of sugar and fats (including fat in meats and dairy products) and against excessive total calorie intake and resulting high body weights. Government agencies in the USA, Canada, the UK, Scandinavia, and other countries are now spreading this advice. And some medical advice would favour also reduced consumption of red meat and eggs. The rising popularity of physical exercise, and 'good food', together with natural appetites, will limit the acceptance of nutritional advice. But even if none were accepted, the annual increase in effective demand for foodstuffs at points of first sale will almost certainly not be more than 1.7 per cent. The 1.7 per cent would result from (i) population increasing at 0.5 per cent a year; (ii) incomes per caput growing at 4 per cent a year; (iii) an income elasticity of demand of 0.3. Thus $101.7 = 100.5 \times (1 + 0.04 \times 0.3)$.

In addition to consumers' demands, however, the total effective demand in industrialized economies includes *the demands of traders and governments for exports*. The great increases in effective demands in low and medium income countries therefore affect the industrialized countries. And the nutritional needs that have not the backing of adequate purchasing power in the poorer countries may acquire this backing from government and private sources of aid, and so add to effective demands for exports. These may be raised also by purchases to meet demands from the USSR and other centrally planned economies (Tables 1.2 and 1.7).

Table 1.7   Average annual increases in foodstuff production[a].

Country groups	Human population 1980	Total foodstuffs			Foodstuffs per caput		
		1963 to 1970	1970 to 1975	1975 to 1980	1963 to 1970	1970 to 1975	1975 to 1980
	per cent of World's	*per cent per year*			*per cent per year*		
Market economies							
Developing[b]	49	3.3	2.7	3.1	0.7	0.3	0.7
Developed	18	2.2	2.1	1.9	1.2	1.2	1.1
Centrally planned economies							
Asian	24	2.7	3.2	3.2	0.9	1.2	1.8
Other	9	2.9	2.6	0.4	1.9	1.7	−0.3
World	100	2.7	2.5	2.2	0.8	0.6	0.5

Sources: FAO (1977a,b, 1982).
[a]Changes in 3-year moving averages of index numbers of foodstuff outputs at the 1969–71 prices of individual countries.
[b]Including capital surplus oil exporters.

1.5   WHAT WERE THE PAST CHANGES IN FOODSTUFF PRODUCTION?

1.5.1   Outputs of agriculture

The FAO calculates, nation by nation, index numbers of the total values of foodstuff production at the prices paid to producers during the base period. The world totals increased by 2.7, 2.5, and 2.2 per cent a year during three periods since 1963 (Table 1.7). In the developing market economies the paces were somewhat faster but still well below the yearly 3.7 per cent recommended in the indicative world plan (IWP) for the whole period 1963–85 to meet a 3.9 per cent increase in demand. In the developed market economies the paces were 2.2, 2.1, and 1.9 per cent. The differences in the rates of increase in different short periods in Table 1.7 are partly due to year to year variations in weather. The differences are greater for annual crops such as cereals than for the total output of foodstuffs including animal products. Because cereals provide about 60 per cent of the calories and 55 per cent of the protein in developing countries, and are the most important group of foodstuffs moving between nations, the year to year changes in cereal production are especially important. Year to year percentage changes are averaged in Table 1.8, column 6, without regard to whether they were increases or decreases, so that better indications are given of insecurities. Of course, within large groupings of various climates, year to year variations in crop yields tend to offset one another statistically. Variations in the Canadian and the Soviet grain crops are comparatively large, because of their heavy dependance on 'prairie' climates (Table 1.8, footnote c).

Table 1.8   Production of all foodstuffs and of cereals, 1966–80.

1	2	3	4	5	6
Types of economy	All foodstuffs, average annual		Cereals, 1969 to 1971	Cereals, average annual^c	
	Net increase^a	Change^b		Net increase^a	Change^b
	per cent		*million tonnes*	*per cent*	
Market economies					
Developing	2.7	2.7	363	2.6	4.1
Developed	2.1	2.3	423	3.0	5.5
Centrally planned economies					
Asian	2.8	2.9	228	3.3	4.7
Other	2.8	5.5	232	3.5	14.2
World	2.3	2.5	1245	2.7	4.0

Sources: Based on FAO (1981, 1982).
^{a}Simple arithmetic means of year to year changes in percentages.
^{b}Averaged without regard to sign.
^{c}For India the figures were 3.0 and 8.3; for Canada, 3.7 and 14.8; for the USSR 6.6 and 22.1.

1.5.2   Ways to improved outputs

Thus both trends and fluctuations about trends pose problems for the future of production, trade, and international relations. Understanding of past changes is therefore needed. The ways in which trends have been achieved, and some fluctuations controlled, may be classified as shown below. Examples of important international data are summarized in Tables 1.9 and 1.10.

Table 1.9   Land use and some major aspects of crop production in 1979, and changes from 1970 to 1979.

	Market economies		Centrally planned economies		World
	Develop-ing	Developed	Asian	Other	
1979					
			million hectares		
Arable and permanent crops	667	392	112	278	1449
Total[a]	4266	2160	625	1616	8667
			per cent of total		
Arable and permanent crops	16	18	18	17	17
Permanent grazings	35	41	56	24	36
Woodlands and forests	49	41	26	59	47
		hectares per 100 ha of arable and permanent crops			
Irrigated area, 1979	15	8	47	8	14
		kilograms per hectare of arable and permanent crops			
Nitrogen in inorganic fertilizers, 1979	17	58	n.a.	43	40[b]
		number per 1000 ha of arable and permanent crops			
Tractors, 1979	3	35	6	13	14
			kilograms per hectare		
Crop yields, 1978–80:					
Wheat, grain	1426	2375	1806	1895	1885
Rice, paddy	2110	5571	3744	3797	2703
Sorghum, grain	832	3220	2341	1280	1370
Changes from 1969–70 to 1979					
			per cent increase per year		
Arable and permanent crop area	0.7	0	−0.2	−0.1	0.3
Irrigated area	2.1	0.9	2.2	4.9	2.2
N in inorganic fertilizers	8.2	4.3	n.a.	5.2	6.1[b]
Tractors	6.9	2.1	12.3	3.3	3.0
Yields					
Wheat	2.5	1.2	4.8	2.2	2.2
Rice	1.5	0.4	2.2	1.1	1.6
Sorghum	2.8	0.3	4.3[b]	−1.2	1.9

Sources: FAO (1981, 1982).
[a]Including permanent grazings, woodlands and forests.
[b]Estimated.

Table 1.10   Numbers and yields of animals: 1979 and changes from 1970 to 1979.

	Market economies		Centrally planned economies		World
	Develop-ing	Developed	Asian	Other	
	per 100 ha of total area[a]				
Numbers, 1979					
Cattle and buffaloes	37	22	17	22	29
Sheep and goats	37	28	41	28	34
Pigs	5	15	69	20	17
	kilograms per year				
Yields, 1979					
Milk per cow	654	3888	661	2338	1916
Beef and veal per head[b]	15	83	31	62	38
Pigs per head[b]	34	116	51	82	70
	per cent increase per year				
Changes in numbers from 1969–70 to 1979					
Cattle and buffaloes	1.2	0.5	−0.6	1.8	1.0
Sheep and goats	0.9	−1.5	1.8	0.5	0.4
Pigs	2.2	1.4	5.0	3.3	3.3
	per cent increase per year				
Changes in yields from 1969–70 to 1979					
Milk per cow	0.6	1.5	3.3	0.8	0.4
Beef and veal per head	0	0.9	2.8	0.9	0.8
Pigs per head	1.3	1.5	0.1	−0.4	0.3

Sources: FAO (1981, 1982).
[a]Excluding woodlands and forests.
[b]In whole cattle or pig populations, respectively.

1.00   *Changes in land use, and land improvements*
1.01   Increasing the areas for sown or planted crops and reducing the areas of grassland, rough grazings, bush, jungle, etc.
1.02   Reclaiming land for cropping from swamps, sea, etc.
1.03   Draining and flood control
1.04   Irrigation, and improvements in irrigation
1.05   Improving soil fertility in other ways

2.00   *Higher yields of particular crops*
2.01   Using improved varieties, suited to local climatic, soil, and other conditions
2.02   Improving practices in the use of fertilizers

2.03 Improving practices in control of weeds, insects, fungi, viruses, and other biotic factors

2.04 Ensuring good physical conditions for crops, particularly timely planting, weed control, harvesting, and storage

3.00 *Improved cropping plans – choices of species used, and the ratios between them, and sequences*

3.01 Suiting local conditions for plant growth – climate and soils, pests, yield possibilities

3.02 Suiting local labour, skills, and mechanization possibilities

3.03 Suiting market and price conditions, as well as household consumption needs and preferences

3.04 Improving rotations, and, where appropriate, mixed cropping and multiple cropping so that more than one crop is harvested per year

3.05 Improving the integration of cropping and livestock plans

4.00 *Increased numbers of ruminant animals (cattle, sheep, etc.)*

4.01 Improving perennial grazings, and rotational (those at times resown)

4.02 Improving feedingstuffs supplies for seasonal and other shortage periods

4.03 Having larger animal numbers

5.00 *Improved stocking plans and management, ruminant animals*

5.01 Relating numbers better to grazing and other feedingstuff supplies

5.02 Choosing better ratios between species, and ages within species, and grazing plans

5.03 Using improved types within species

5.04 Improving the integration of animal with cropping plans

6.00 *Expanded and improved pig, fowl, and other non-ruminant production*

6.01 Attending to housing, mating, and care at birth and of the very young

6.02 Improving the integration of animal with cropping plans

6.03 Improving nutrition

6.04 Using improved types – breeds and races

6.05 Fish farming

7.00 *Control of animal diseases and pests*

7.01 Inoculating and otherwise preventing diseases

7.02 Controlling movements, and compulsory slaughter, of animals with some diseases

7.03 Hygienic housing and managing to reduce stresses and diseases

Even such a brief listing of ways shows them to be numerous and to require understanding of many determinants of success or failure along them.

1.6 WHAT ARE THE OBSTACLES TO FUTURE INCREASES IN FOODSTUFF PRODUCTION?

In low and medium income countries many obstacles must be overcome. Obstacles that have slowed increases in production during recent decades and are still important can be briefly listed:

(a) Lack of *detailed* scientific knowledge, *locality by locality*, of climates, soils, biotic factors, crop plant materials, and farm animals

(b) Lack of *detailed* knowledge of markets and prices, *locality by locality*

(c) Lack of ability and energy on farms to plan greater production increases, and bear their risks and uncertainties and managerial commitments

(d) Lack of more favourable prices, market outlets, and taxation conditions

(e) Lack of capital (assets for production) and loans at suitable interest rates

(f) Lack of land (and related water) resources, unfavourable patterns of distribution of land-use rights, and social and other obstacles to land-use changes (other than (a) above)

(g) Social and other obstacles to migrations of labour and re-settlements to secure distributions of labour more nearly optional for production

(h) The additional manual labour required that is not forthcoming from labour forces or is lacking because of poor food supplies and ill health; and all the reasons, in design, cost, and distribution, why improved tools, and other equipment are not used to meet the requirements

In the industrialized economies the same list can be used, although with many changes between and within countries. A general summary would indicate the following:

	Least important	⟶		Most important
North America	(e), (f), (g), (h)	(a), (b), (c)		(d)
Western Europe	(e), (g), (h)	(a), (b)	(c), (f)	(d)
Australia	(f), (h)	(c), (e), (g)	(a), (b)	(d)

Now for the world as a whole special note should be taken of the following obstacles:

(i) The need properly to identify the land that should be 'taken into' cropping systems (and the land that should be 'taken out')

(ii) The social, capital, and other obstacles to 'taking-in' land, and the servicing of it with roads, towns, etc.

(iii) Rural populations where these are already, or soon will be, much higher than farming plans require, and where land is not available to 'take in'

(iv)   Lack of resources in research and education for further increases in crop yields and biological and economic improvement of whole farm production plans, *locality by locality*

(v)   The oil and other energy requirements (including those for fertilizer production) of expanding crop areas and raising crop yields

(vi)   Problems of many kinds in control of pests and diseases

(vii) The desirability everywhere of greater conservation of natural resources and better defences against pollution and against hazards to health of humans, crops, and animals

(viii) Lack of purchasing power in the low income countries and particularly in their poorest groups, and slow increases in effective demands for food in the industrialized countries

(ix)   The various constraints on international trade, and loans and investments.

1.7   CAN FOODSTUFF SUPPLIES BE MORE SECURE FROM YEAR TO YEAR, AND PRICES AND PRODUCERS' INCOMES MORE STABLE?

We have noted that variations in production are important and need to be reduced or offset if food supplies are to be secure. Attempts to make crop production less variable raise many biological, engineering, and economic problems. These increase when the frontiers of cropping are pushed farther into less reliable climates and on to poor and less well understood soils. Economic questions arise that require some political decisions (e.g. about international trading, international aid, and management of costly reserve stocks). Variations in livestock production result from variations in yields of feedingstuffs, including grazings, and from farmers' decisions. Attempts to reduce these variations and avoid serious deficits and surpluses also raise many and diverse problems, some requiring political decisions (e.g. about milk production in western Europe and disposal of surpluses as aid to poorer countries).

A principal reason for 'political' problems is that prices affect decisions about what, where, and how much production is intended, and about trading and stock holding. Prices also affect producers' incomes, and costs to consumers. But price relationships are many. And they have to change often, so as to relate changes in consumption, trade, and storage to the changes in supplies due to past decisions and to weather, pests, and other natural causes. Therefore wise decisions by producers, traders, stockholders, and governments require adequate understanding of the functions of prices and of the various influences on future prices. For such decisions research, education, and objective political discussion are needed now more than ever.

1.8   SHOULD TRENDS IN AGRICULTURAL PRODUCTION OF RAW MATERIALS BE CHANGED?

This is an important question because many nations want to earn foreign exchange by production and export, and many want to import raw materials to sustain industry and meet consumers' demands.

The raw materials produced, and the biological and economic conditions for production, are very varied, as Table 1.11 indicates. Production possibilities and how far these have been grasped vary widely. Also demands change, because the demands for the final products fluctuate with levels of general economic activity and availability of substitutes, and alter with consumers' preferences. Thus the trends in production and in 'real' prices for particular raw materials show significant contrasts (e.g. between those for rubber and sisal; cotton and wool; cocoa and tea in Table 1.11).

Table 1.11   Production and prices of some raw materials.

	Coffee, green	Cocoa, beans	Tea	Cotton, lint	Sisal	Wool, greasy	Tobacco	Rubber
	million metric tons							
Production								
1961–5	4.4	1.2	1.1	10.9	0.6	2.6	4.4	2.2
1971–5	4.5	1.5	1.5	13.3	0.7	2.6	5.0	3.2
1976	3.6	1.4	1.7	12.2	0.4	2.6	5.7	3.6
1977	4.3	1.4	1.8	14.2	0.5	2.6	5.7	3.6
1978	4.7	1.5	1.8	13.2	0.4	2.7	5.8	3.7
1979	5.0	1.6	1.8	14.0	0.4	2.7	5.4	3.9
1980	4.7	1.6	1.9	13.9	0.5	2.8	5.2	3.8
1981	5.8	1.7	1.8	15.3	0.5	2.8	5.3	3.8
	index numbers: 1975 = 100							
Prices[a]								
1961–5	140	54	175	90	83	170	108	172
1971–5	108	99	107	114	106	134	112	117
1976	205	150	105	127	77	141	105	133
1977	327	287	178	107	79	145	102	128
1978	205	245	132	100	68	139	104	143
1979	199	169	117	96	91	143	98	169
1980	156	119	106	108	86	125	96	171

Sources: FAO (1981, 1982).
[a]Prices based on series of international significance, in real terms, using US Department of Labour index of producer prices (all commodities) as deflator.

1.9   WHAT ARE EXAMPLES OF PROBLEMS IN SECURING OTHER SOCIAL BENEFITS IN THE FUTURE?

1.9.1   Youth

In rural areas many of those who are about 15–22 years old face special difficulties in choosing wisely their future occupations. They have little or no knowledge of alternatives to agriculture and of economic trends, limited training for possible transfer to other locations, inadequate knowledge of social conditions there, and little or no money. The problems in trying to overcome these difficulties by providing information and education are substantial.

There are also difficulties in fostering in urban youth (i) wise decisions about possible future occupations in rural areas, and (ii) understandings of possible recreational and educational benefits obtainable from natural endowments of the land.

1.9.2 Conservation

For these benefits, and for future crop and animal production, conservation is required of soil and water resources and the natural habitats of wild species. Complex problems therefore arise that need for their solution knowledge, foresight, and management. Some also require political valuations of recreational and educational benefits, as against foodstuff and other agricultural products (see Section 21.8).

1.9.3 Social divisions

In many poor countries social problems run still deeper. Relations between rural and urban groups are unsatisfactory because of what has been called 'urban bias' in pricing, trade, and taxation policies. Where land and capital are in short supply because of population pressures, the mass of farmers judge that many landowners and money lenders gain too much both socially and economically. And where modern developments in use of machinery, irrigation, and chemicals are promoted, those with most land, capital, and education often gain most. In some areas their gains are at the expense of the poor farmers so that social divisions are aggravated.

1.9.4 Numbers and sizes of farms

In some industrialized economies, and in the centrally planned ones, political and bureaucratic ideas about the benefits of fuller mechanization of agriculture and larger farms may also raise serious questions. From the standpoint of social relationships and welfare: (i) What should be the numbers of farms and farm families? (ii) What should be the distribution of farms by area or other measures of 'size'? (iii) Who should provide the management and enterprise? And, not least important, (iv) What should be the speed of change? Although such questions have arisen many times in history, and acutely in Russia and China, they are still with us. In the richer countries they are raised, in ways not known a few decades ago, by taxation policies, inflation, and the interventions of governments in providing subsidies for amalgamations.

24

REFERENCES

Åstrand, P. O. (1973). Nutrition and physical performance. *World Review of Nutrition and Dietetics*, **16**, 59–79.
FAO (1967). *Agricultural Commodities – Projections for 1975 and 1985*, Vol. II, FAO, Rome.
FAO (1977a). *The Fourth World Food Survey*, FAO, Rome.
FAO (1977b). *Production Yearbook*, FAO, Rome.
FAO (1981). *Production Yearbook*, FAO, Rome.
FAO (1982). *Production Yearbook*, FAO, Rome.
IFPRI (1976). *Meeting Food Needs in the Developing World*, Research Report No. 1, International Food Policy Research Institute, Washington, DC.
Jelliffe, D. B. (1973). Nutrition in early childhood. *World Review of Nutrition and Dietetics*, **16**, 1–21.
World Bank (1980). *Development Report*, World Bank, Washington DC.
World Bank (1981). *Development Report*, World Bank, Washington DC.

FURTHER READING

FAO (annual). *State of Food and Agriculture*, FAO, Rome.
Krebs, A. H. (1973). *Agriculture in Our Lives*, 3rd edn, Interstate Printers and Publishers, Danville, Ill.
World Bank (annual). *Development Report*, World Bank, Washington, DC.

QUESTIONS AND EXERCISES

1. List and briefly explain the main types of contribution to human welfare from the agriculture of the rural area you know best.
2. Discuss the significance for farmers, consumers, and governments of the low income elasticity of demand for many foodstuffs in the richer countries.
3. During the last ten years in the area or nation you know best, what were the main changes in agricultural production?
4. Find the data for your nation on the following: human population; gross domestic product; total agricultural production; and agriculture in the economic structure. Give the figures for a recent year and past rates of change.

Chapter 2

The purposes of a systematic approach

2.1   AGRICULTURE AS A SUBJECT

Our brief review of the nine central questions of Chapter 1 shows that the answers to them require contributions from various bodies of knowledge. We commonly call these *subjects* – chemistry, biochemistry, physics, climatology, botany, veterinary science, engineering, demography, nutrition, economics, and many more.

Our brief view also indicates that many agricultural problems concern *changes*. The possibilities of *technical changes* in particular directions are commonly found before the desirable technical changes in support are ready for adoption. For example, tractors were available to farmers before suitable machinery to replace the horse-drawn. New high yielding rice varieties (HYV) are available to suit the climates of some areas where all relevant detailed local knowledge is not available about irrigation possibilities, soils, fertilizers, and pests. Agricultural problems with solutions requiring *economic and social changes* are liable to be regarded as political problems about power in governments, and conflicts of interest between particular groups – rural and urban; rich and poor; landowners and tenants; the industrialized 'North' and the agrarian 'South'; and so on. Also, because agricultural production practices differ between *areas and regions*, those who have experience in only one or two localities find difficulties in grasping easily the different priorities and practices desirable elsewhere. Proposals for changes need to be carefully formulated and tested, but the costs of research and development work are high, and years are often required. Narrow judgements are therefore commonly made, based on fear of changes. They slow the pace of change.

For all these reasons the difficulties are substantial in studying *agriculture* as a subject. The scope and complexity of its purposes are greater than those of medicine. Methods of approach are therefore especially important.

2.2   THREE APPROACHES TO THE STUDY OF AGRICULTURE

A common approach is to consider particular *observations and developments* and

25

try to relate them to local knowledge and practices. The uses of bones, lime, and organic manures to improve soil fertility were discovered and widely adopted long before the scientific reasons were found. Many local 'classifications' of soils in Africa and elsewhere in the Tropics are based on local observations of the natural vegetation in them, the work required to till them, and the yields of crops they can grow. The 'horse-back guesses' of possible crop yields in the opening up of much of the Mid-West and Great Plains of the USA were based on the relation of the species and height of the natural vegetation to soil fertility. Many of the species we use as crop plants have been spread from their area of origin, and much improved, by little more than close local observations and care.

Another approach is to consider the *problems of particular regions or areas*, and to try to reach satisfactory solutions to them. In other words, attempts can be made to elaborate the nine questions of Chapter 1 for more local use. For example, in a particular tropical country where the human population is increasing at 3 per cent a year but the real value of exports of tea and other agricultural products is tending to fall, how can rice production be increased? What priorities should be given to other grains, or starchy roots, as against rice? What new production techniques would be feasible and economic? Should other agricultural changes be given priorities? Or mining, or industrialization? And so on.

A third approach has advantages over the first and second in that it can use knowledge from all pertinent sources and integrate it well for solutions of the problems of all areas, or show clearly what important types of knowledge are missing for particular areas. Narrow views, false conclusions, and missed opportunities can therefore be avoided. This third approach requires a basic classification of knowledge from the various specialist subjects and from farmers and others with local experiences and responsibilities. The system of classification is designed as that most useful when tackling detailed elaborations for particular areas of the nine questions of Chapter 1. The approach is thus based on a *framework for systematic study*, to help actual decision making.

Agriculture is an integration of knowledge of many sorts. This third approach discloses its nature best.

2.3 AN EXAMPLE OF THE NEED FOR SYSTEMATIC STUDY

An example that was important in western Europe in the 1970s can be given briefly. The marketing arrangements and prices of milk indicated to farmers in an area of mixed farming that they could sell more milk. They could produce it by keeping more cows and/or raising yields per cow. To do so they would have to feed more grain and/or more grass and hay and/or grass silage. They could also improve health and reproductive efficiency in cows. What future prices should they expect for milk and for grain, complementary protein feedingstuffs, nitrogen fertilizers, machinery use, and other inputs? What varieties of grass should they use, and in what mixtures? How much nitrogen should they use? What other fertilizers, and how much? Would there be fungus diseases and frost damage in the grass? Would there be pollution dangers from additional fertilizers, or silage juice? How should they manage the cattle on the grazings? When and how should they cut grass for hay or silage? Would there be deficiencies of minerals to be made good in the cows' diets? Or actions to be taken to protect the health of calves? How should the additional work be carried out? What combinations of labour, machinery, equipment, and building should be used? What would the effects of alternative plans on capital requirements and on farm expenditures and income be? How long could increasing supplies of milk be sold at the expected prices? How will prices of land, rents, wages, machinery, fuel, fertilizers, feedingstuffs, etc., alter in the future? Will there be new legislation about milk quality, or international trade, or the employment of labour? What differences would the new grassland management make to other crops? How would the additional cow faeces and urine be used? Would small grain crops lodge before harvest and so be damaged, because of changes in soil fertility? Would the costs of growing potatoes or vegetables be reduced by these changes?

Much of the thinking has to be 'horizontal' – from one 'subject' to another. This example starts in markets and pricing arrangements for milk. It carries us to cow nutrition, and veterinary, botanical, and other biological questions, including those of soil science and pollution. Then to questions about labour tasks and equipment; and on to those of farm production economics and finance. Then we return to markets, international trade, consumer protection, pollution control, and other possible socioeconomic issues. And back again to biological questions.

2.4   THE FOUR SUB-SYSTEMS OF AGRICULTURE

The example indicates that our framework for systematic study, and our conception of agriculture as a subject, can best be based on the idea that there are four sub-systems in agriculture:

(a)   *Biological* – integrating knowledge about plants and animals and about the biological effects of physical and chemical matters (climates, soils, etc.) and of man's activities (drainage, irrigation, use of fertilizers, tillage, artificial insemination, vaccines, etc.) (Chapters 3–6)

(b)   *Work* – integrating knowledge about the physical tasks in agriculture, and about how they can be well completed by combinations of labour and skills, tools, machinery, buildings, and other equipment, with energy from various sources (Chapters 7–9)

(c)   *Farm production economics* – integrating knowledge about prices of products sold and factors of production bought by farmers; quantities produced; quantities used; management and labour 'costs', receipts, and expenditures; net surpluses; alternative production plans; risks and uncertainties; and all other determinants of farm incomes and detailed production systems (Chapters 10–12)

(d)   *Socioeconomic system, those parts that affect (a), (b), and (c)* –

integrating knowledge about markets for farm products (including international trading); markets for land and land-use rights; labour; machinery, fuel, fertilizers, and other factors of production; loans; and about taxation and subsidization; research; education (Chapters 13–20).

We can recognize without difficulty that each sub-system affects all three others. In any one area the physical tasks are related closely to the biological conditions. The farm-economic calculations are meaningless if they are not based on the biological and work sub-systems as well as on the socioeconomic. And the socioeconomic system has been much affected by biological, work, and farm-economic conditions in the past.

All four sub-systems can be quickly understood by specialists in single subjects in any one sub-system, at least sufficiently to provide real help, and encourage effective liaison throughout all four.

Such a framework for systematic study is therefore well suited to the needs of those who are at early stages in the study of agriculture, and of those who have come to it by specialist routes, or political or administrative responsibilities.

Unfortunately the words 'system' and 'sub-system' are often confused with words for mathematical models that are designed to simulate, sufficiently well for chosen purposes, selected biological or economic and biological interrelationships. The sub-systems discussed in Parts II–V are not such mathematical models. Their purpose is to help more directly with 'horizontal thinking', conception of interrelationships, and choice of scope and purposes in data collection. These are always highly desirable, and when useful mathematical models are to be designed they are essential.

Another important benefit from our systematic framework is help in defining '*development*' in a way that has wide validity. Likewise an adequate *conception of* '*policy*' in relation to agriculture can be built up (Chapter 21) and some guidance secured on methods of carrying policies into effect (Chapters 22 and 23).

2.5 RELATION OF STUDIES TO FARMS

Because large numbers of farms are each responsible for their own agricultural production activities, many different aspects of the four sub-systems can be illustrated by data for farms that have different biological, work, farm-economic, and socioeconomic conditions.

Within the scope of this book, it is possible to summarize and use data for nine groups of farms that have very different conditions. But readers will find it useful to make similar summaries for farms in the areas that interest them most.

Table 2.1 lists the areas from which the example farm groups were drawn, and indicates their locations, climates, and soils. The purpose of the selection is to provide contrasts. The averages for the example groups are not necessarily precisely representative of whole areas. Thus the central Illinois and Saskatchewan groups specialized in cash crops and have almost no animals,

Table 2.1   Locations, climates, and soils of example farm areas.

Areas	Main sources of data	Location (approx.)		Climate classes[a]	Soil orders and sub-orders[b]
		°N	°W/°E		
Aberdeen, Scotland, UK	Isaacs (1981)	57	2W	Sr 21 1	A1a
Central Norfolk, UK	Murphy (1980)	53	1E	Sr 21 2	A2d
Cortland, NY, USA	Knoblauch (1980)	43	76W	Hr 32 3	A2a
Central Illinois, USA	Wilken et al. (1979)	41	89W	Hr 32 3	M4a
Southern Saskatchewan, Canada	Saskatchewan Agriculture (1979)	52	106W	Dd 22 3	M2a
South-western Louisiana, USA	Johnson et al. (1978)	30	92W	Hr 30	A3d
Iloilo, Philippines	Barlow et al. (1983)	11	123E	Hrl 30	X3
Aurepalle, AP, India	Ryan (1982)	18	78E	Dwm 30	A3j
Western Malaysia	Barlow (1978)	3	102E	Wr 30	U3g

[a]The first letter indicates the 'precipitation effectiveness', according to Thornthwaite (as stated in Janick and others, 1974) as the sum of the monthly precipitations divided by respective monthly evaporations:

W   is wet, sum 128 or more;   S   is sub-humid, sum 32–63;
H   is humid, sum 64–127;   D   is semi-arid, sum 16–31.

The second letter indicates the seasonal distribution of precipitation:

r   is abundant at all seasons;   s   is sparse in summer;
d   is sparse at all seasons;   w   is sparse in winter.

A third letter may indicate the number of months that on average have more than 50 mm of precipitation:

l   is 8–11 months   m   is 5–7 months

Average temperatures are indicated by the first number for the warmest month and by the second number for the coldest:

	Warmest month		Coldest month
Over 20°C	3	Over 13°C	0
10–20°C	2	2–13°C	1
6–10°C	1	Under 2°C	2
Under 6°C	0		

The range of temperatures between the coldest and warmest months averages is indicated by the third number (in some areas only):

36–48°C	4	12–24°C	2
24–36°C	3	Under 12°C	1

The third letter and the numbers are based on the classification system of D. L. Linton as set out in the Oxford Atlas of the World.

[b]Based on the classification system of the US Department of Agriculture.

although within their areas some farms produce cattle and pigs. The south-western Louisiana group data are for a completely modernized production system.

The summary tables are Tables 3.2–3.6, 9.5, 10.1 and 10.2, and 12.2. Other information for the example farm groups is used in the text for illustrative purposes.

REFERENCES

Barlow, C. (1978). *The Natural Rubber Industry*, Oxford University Press, Kuala Lumpur.

*Barlow, C., Jayasuriya, S., and Price, E. C. (1983). *Evaluating Technology for New Farming Systems: Case Studies from Philippine Rice Farming*, International Rice Research Institute, Los Banos, Philippines.

Isaacs, R. J. (1981). *Farm incomes in the north of Scotland 1976–7 to 1978–9*. Financial Report 77, School of Agriculture, Aberdeen.

Janick, J., Schery, R. W., Woods, F. W., and Ruttan, V. M. (1974). *Plant Science: An Introduction to World Crops*, 2nd ed., Freeman, San Francisco, pp. 265–266.

*Johnson, B., Bordelon, K., and Heagler, A. (1978). *Capital requirements and incomes for major crop systems in four agricultural areas in Louisiana, 1976*. Department of Agricultural Economics Research Report No. 536, Louisiana State University, Baton Rouge.

Knoblauch, W. A. (1980). *Dairy farm business summary: central New York*. Agricultural Economics Ext. 80–15, New York State College of Agriculture, Ithaca, NY.

Murphy, M. C. (1980). *Report on farming in the eastern counties of England, 1978–9*. Agricultural Economics Unit, Cambridge University, Cambridge.

Ryan, J. G. (1982). Unpublished data from the International Crop Research Institute for the Semi-Arid Tropics, Patancheru, AP, India.

Saskatchewan Agriculture (1979). *Farm Business Review for the Year 1978*. Statistics Branch, Saskatchewan Agriculture, Regina.

Wilken, D. F., and others (1979). *Summary of Illinois Farm Business Records: 54th Annual, 1978*, University of Illinois, Urbana-Champaign, Ill.

FURTHER READING

*Bayliss-Smith, T. P. (1982). *The Ecology of Agricultural Systems*, Cambridge University Press, Cambridge.

Duckham, A. N., and Masefield, G. B. (1970). *Farming Systems of the World*, Chatto and Windus, London.

Grigg, D. B. (1974). *The Agricultural Systems of the World: An Evolutionary Approach*, Cambridge University Press, London and New York.

EXERCISES

1. Choose one of the main changes in agricultural production during the last 10 years in the area or nation you know best. Under the headings 'biological', 'work', 'farm-economic', and 'socioeconomic', list and briefly explain the nature of this change.
2. Briefly contrast agricultural production in two different areas or nations, under the headings 'biological', 'work', 'farm-economic', and 'socioeconomic'.

Part II

The biological sub-system

Chapter 3

Introduction

Mankind depends on many species of plant and animal. The methods of securing their products vary from gathering, hunting, and range management to intensive horticulture and even soil-less culture (hydroponics). But all methods depend on the sound functioning of the species in growth, development, and reproduction. And almost all methods alter the dynamic balances of Nature and so result in pressures to re-assert these balances by weeds, pests, and disease organisms – the biotic factors.

3.1 SUMMARY OF FLOWS

3.1.1 Introduction

The simplest framework for our systematic studies is illustrated in Figure 3.1. The contents of the 'boxes' in this figure, and the flows into and out of them, are considered briefly in this chapter. Reference back to the figure will help when each section is read. At the end, special note is taken of the biotic factors, hydroponics, and fish farming. The importance of interdependence is then noted.

The flows in Figure 3.1 include all forms of energy and all chemical elements and compounds. The lines represent flows in both directions, except where arrows are shown (e.g. plants use light and carbon dioxide from climates and atmospheres, but also affect atmospheres by transpiring water, shading sunlight from other plants, and respiring carbon dioxide). No attempt is made to show controls of flows.

3.1.2 Humans

The outputs of the biological flows that are used by humans as food and raw materials come from different parts of many species of plants, and from a smaller number of animal species. Plant product examples are: wheat seed; potato tubers; sugarcane stalks; rubber tree latex. Animal product examples are: milk; meat; eggs; wool; hides; silk. From some of the foodstuffs from which human foods are taken, the *by-products* are used as industrial raw materials (e.g. glands from cattle for pharmaceutical industries).

All outputs vary widely in *quality*. For example, wheat seeds (grains) vary in

34

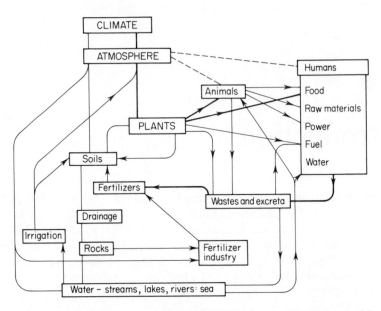

Figure 3.1   The biological sub-system of agriculture: A simple diagram of flows.

size; shape; colour; content of water, gluten, starch, sugar; amount of fungus infection; and in other ways affecting the quantity and quality of the flours that will result from milling the grains. Many different factors (including flavours) go to make up what consumers value as quality (e.g. in rice, potatoes, tea). Cotton lints vary in length, and fineness; cotton seeds in oil content. Apples may be damaged by fungus diseases. Milk varies in content of fat, and 'non-fat solids'. Beef and other meats vary in percentage contents of lean flesh, fat, and bone; tenderness; succulence; flavour. Milk, meat, and eggs may contain bacterial and other diseases (e.g. tuberculosis, brucellosis, salmonellosis).

The outputs of animal *power* are large from some systems (e.g. bullock power in India). The outputs of *fuel* are also important. They include straws; dried faeces; twigs and other tree products; methane gas from decaying plant residues and faeces.

Water is included in the *humans* box because the availability of clean drinking water is essential to health, and because systems of agriculture can affect supplies.

The kinds and qualities of output to *humans* differ widely from place to place. Decisions about *what to produce* depend on (a) human preferences; (b) natural endowments of climates and soils in relation to human populations; and (c) human skills and energies in using these endowments. Preferences are based on human physiology, but also on habits founded on past experiences of supplies and uses of foodstuffs and raw materials.

One very important type of decision determines the flows along the vegetarian (plants–humans) pathway as against flows along the plants–animals–humans pathway. Animals require much energy for their basic metabolism, body repairs, movements in grazing and feeding, etc., so much energy is 'lost' in the conversion of energy in feedingstuffs into energy in milk and other animal products.

But the plant products of some climates and soils are not suitable for direct human use. For example, where climates are too cold or dry or wet, or soils too thin, tillage crop outputs require too much human effort. The vegetation grown is managed only as ranges or rough grazings, producing high-fibre material suited as feedingstuffs only for sheep, goats, cattle, reindeer, or other livestock capable of digesting it (as in Masai country in Africa; Wyoming, USA; the Highlands of Scotland; Lapland).

Another reason for animal production is that some of the by-products of foodstuff processing are not suitable or preferred as human foods, and so are used as feedingstuffs (e.g. meals from oil seeds after extraction of oil; blood and bone meals from slaughterhouses). Also kitchen wastes, and low quality grains, damaged tubers, and spillages on farms are important as feedingstuffs for poultry (e.g. in India and Africa), or poultry and pigs (e.g. in China).

The flows out of the 'humans' box, *processing and kitchen wastes and faeces and urine*, are biologically important flows to the 'wastes and excreta' box. In modern towns these are usually treated as refuse and sewage and not returned to soils as fertilizers. Some go to sewage farms. Elsewhere 'night-soil' is very important in the maintenance of soil fertility (e.g. in China, around Kano in northern Nigeria; in household plots in most tropical countries). Blood, bones, and other by-products from the slaughtering of animals are commonly partly used as fertilizers.

It is also important to note that some industrial and other wastes, as well as excreta, can pollute water and soil resources.

3.1.3 Plants

Crop plants (including those in grazings) can usefully be regarded as central. They use the flow resources of water, heat, light, and carbon dioxide from 'climates' and 'atmospheres' and those of nitrogen, phosphorus, and other nutrient elements from 'soils'. They synthesize sugars, starches and other carbohydrates, amino acids and proteins, and all the other compounds required for plant tissues and reserves, including all the variety of 'parts' that we secure as outputs. The conditions for growth and development are therefore to be found in 'climates' and 'atmospheres', and 'soils'. But these differ widely from place to place, and the flows from them vary seasonally and from year to year. How then can plants make best use of all this variety?

In Nature the answer is to have a great number of plant species. Any one species is so suited to particular environments that it survives in them. In its *niche* in the ecological pattern it is better suited than most other plant species to the climate and soil, including the seasonal changes in these, and to the numbers and types of other living organisms. And the suiting is inherited. But variations do occur, and past variations have provided over the centuries opportunities for natural selection of the great number of species now living. To secure a useful understanding, therefore, we need to consider the following: (a) types of difference between plants that suit them well to particular environmental conditions; (b) how such differences occur and are inherited.

In agriculture, environments for crop plants may be altered by irrigation, tillage, weed control, and other methods. And the aims in agriculture are not for survival alone but also for high and reliable yields and qualities. We must therefore consider also (c) how much differences in yield (and quality) are related to (i) inherent differences in plants that suit them to particular, perhaps improved, environments and (ii) differences and improvements in environments, for different stages in plant life histories. The ways in which differences in plants are inherited are obviously important to breeders and selectors. How these differences actually control crop plant yields are important (i) in defining the 'ideal plants' that breeders should aim to obtain and (ii) in understanding and improving crop production (agronomic) practices.

The outflows from *plants* include feedingstuffs for *animals* and foodstuffs, raw materials, and fuels directly for *humans*. Two other types of outflow are (i) straw and other by-products to *wastes and excreta* to be used eventually on selected sites as soil fertilizers; (ii) plant roots, straw stubbles, and other parts that remain in the fields where grown, and are broken down there by fungi, bacteria, and other organisms, so 'flowing' back to affect the fertility of the soil.

To keep the framework simple, special boxes are not separately shown in Figure 3.1 for the seed and other outputs of crop plants used to start future production. In practice, how these are selected, secured, stored, and distributed can greatly affect outputs (Section 17.3.2).

3.1.4 Climates and atmospheres

The differences from site to site are so great that the plant and animal production required by mankind cannot be secured without *intricate patterns* of (i) many crop plant species and varieties (cultivars); (ii) agronomic practices (e.g. about sowing and harvesting dates, soil management, crop plant sequences).

The major purpose of the patterns is to provide, at the right times and at the right stages of growth and development, the species and cultivars best suited to the environments of particular sites. Much therefore depends on choices of cultivars and timings.

The flows from climates to animals are not required for photosynthesis. The relations of animals and humans to climates are therefore not so direct and intimate as those of plants to climates. The geographical distribution of animals and humans is largely the result of plant–climate patterns and animal and human dependence on plants. But such dependence has resulted in the evolution of some animals directly affected by climatic factors; their controls of reproduction depend on day lengths. Thus under natural conditions lambs are born and chickens hatched at the seasons most favourable for their growth and development. Other environmental needs do pose some problems in animal production (Section 5.5.1).

The effects of *climates* on *soils* and *water* are of major importance. Soil water on which plants depend comes from rain and snow. If rainwater falls too fast, some of it may run off the surface of the soil, carrying soil particles with it and so causing *soil erosion*. Of the water that can permeate the soil, only some is retained for use by plant roots. Some drains away, possibly carrying nitrates and other plant nutrients in solution. Climates can also dry soils, and if soil water contains salts in solution these may be left in the soil in damaging concentrations.

Also to be noted are the effects that temperatures, water, ice, and wind have had on soil formation and soil characteristics.

3.1.5 Soils

The chemical, physical, and biological conditions in soils for plant roots differ greatly from site to site. The reasons are mainly that, as affecting soil formation: (i) parent rocks differed widely in chemical composition; (ii) climates differed widely; (iii) topographies affected the movements of water, stones, and soils. And, as affecting present-day plant–soil relationships, there are also the effects of (iv) the current climates and topographies. The natural endowments of soil resources for plants are therefore spread over the Earth's surface in very complex and intricate patterns. These patterns require careful assessment, along with climates, when deciding what crops to grow on any one site and how to grow them. Some soils suit well many different crop species (e.g. the most fertile soils of the Mid West of the USA). Some suit only certain groups (e.g. soils over chalk rocks in southern England). Some suit only species in the vegetation natural to them (e.g. the wet, acid soils of north-western Europe; soils with a high content of salts in Nevada, USA).

Decisions about what crops to grow depend also on what human efforts would be needed on particular soils. The natural vegetative cover has to be killed, and decayed or removed, before tillage crops can be grown, and in some regions this may require much effort (e.g. in tropical rain forests; high savannah bush; grasslands of the Great Plains of North America). The work of tilling soils and harvesting from them (particularly the harvesting of tubers or other underground parts) ranges from 'easy' on sandy soils to 'impossible' at times on soils that dry to concrete-like hardness. Some soils are on slopes that make tillage and harvesting difficult, and soil erosion serious, unless skill and effort are devoted to preventing it. These characteristics all affect *soil management* (Section 4.9), the work sub-system (Chapters 7–9), and the farm-economic sub-system (Chapters 10–12).

Improvements to soil conditions for plant growth can be made in various ways and these are considered in Section 4.9. From Figure 3.1 we should note that the 'flows' into the 'soils' box are from crop residues left on the sites where grown, the 'wastes and excreta' box, the 'fertilizer industry' box, and from the weathering of subsoils and rocks.

Some possible reductions in soil resources should also be noted from Figure 3.1: (i) soil erosion by water, and wind; (ii) the leaching of plant nutrients in drainage water; (iii) the continuous removal of crop plant outputs where inflows do not compensate fully in chemical terms; (iv) the lowering of the organic matter content where organic matter inflows are insufficient; (v) pollutions.

3.1.6 Water, irrigation, and drainage

Climates determine the water precipitated, the water held in atmospheres, and the evapotranspiration rates. Seasonal changes in all these cause different seasonal run-offs from soils, permeations (infiltrations) into soils, drainage (percolation) through soils, and perhaps slow rises by capillary action in soils. The water available in soils for plant use, and the water held too tightly by soils for plant use, can be regarded as net results. Plant covers and agronomic and other practices affecting them help to determine evapotranspiration, run-off, permeation, and drainage rates, and therefore net changes in underground water supplies. Rock types and formations under ground, and topographies as affecting streams, rivers, and lakes, also determine these underground supplies and their location.

Improvements in water resources and flows from them for agriculture are of various kinds. Often the first was the collection of water for humans; the second, to make the watering of *animals* easier. Where soils were potentially fertile enough but climates too dry, a third was to use water for *irrigation*. This is further considered in Section 4.9.4, with soil management practices to improve water relations (Section 4.9.6).

Improvements in *drainage* are needed in various circumstances. They may be even more important than irrigation (Section 4.9.4).

Reductions in water resources are caused by excess use for humans, animals, or plants, and by excess evaporation, run-offs, and drainage. Faulty cropping and grazing plans, soil management, irrigation, and drainage contribute to such reductions (Section 4.9.6).

Pollutions also can occur, as Figure 3.1 shows, from wastes and excreta, from drainage, and from run-offs. Water supplies for humans and animals can carry many types of living organism damaging to health. In many countries (e.g. India and Indonesia) water-borne diseases are major causes of high death rates. Other types of infection result where water resources are in such forms as favour high populations of insects or other vectors (e.g. mosquitoes carrying malaria; *Simulium damnosum* causing river blindness (*onchoceriais*); snails carrying *bilharzia*). The drainage of plant nutrients can be particularly damaging to the quality of water supplies in areas where large applications of nitrogen fertilizers are made for intensive crop production but neither crop plants nor soils can retain enough nitrogen. Excessive quantities are leached away into underground water supplies, streams, and lakes. Pollutions by irrigation water that is too salt are serious. If the silt content of irrigation water is too high, accumulations of silt can cause serious damage.

Yet another type of damage due to water resources and their uses is the overgrazing in *semi-arid areas*, where drinking water outlets are not suited in numbers or locations to the seasonally changing populations of grazing animals.

3.1.7 Animals

The content of the *animals* box in Figure 3.1 depends mainly on the flows into it from *plants*. The feedingstuffs produced by plants are very varied in palatability, digestibility, net energy value, and nutrient contents. Those with high fibre contents are too indigestible for pigs and poultry, which have to eat more 'concentrated' feedingstuffs (such as grain or kitchen wastes).

But, as compared with the number of crop plant species, the important species of domesticated animals are few in number. Slowly a few others are being used, such as bison, red deer, salmon, trout. The reasons for the small number of species are that (i) animals do not depend like plants *directly* on climatic resources for energy; (ii) the physiology of domesticated species allows them to live – with a few exceptions such as trout and silkworms – in a wider range of climatic conditions than suits most crop plant species; (iii) selection and breeding within animal species have secured breeds and races that suit particular environmental conditions well; (iv) climatic variations are reduced by housing in winter in higher latitudes, and shading in low.

Flows to 'animals' from 'climates' should include clean *water*. Other flows are considered in Section 5.5.1.

The outputs from 'animals' include foodstuffs, raw materials, and power. Some 'wastes and excreta' may be recycled as animal feedingstuffs (e.g. meat, and bone meal); some dried for use as fuel; but most in the higher latitudes are returned to soils. The fertilizing effects are commonly very important. Under intensive grassland production systems, however, pastures may be fouled during grazing seasons and physical damage caused by 'wastes' (sludge) spread in winter.

Animals have important effects on plants also through their grazing activities, particularly (i) where their numbers are too small to 'harvest' outputs of pastures early enough, or (ii) where numbers are too great. Overstocking removes too much plant tissue and reduces plant production and plant cover so that soil surfaces harden and run-off of rains and soil erosion are increased. Where there is selective feeding (as often by sheep and goats), the proportions of the more desirable species in pastures and rough grazings are reduced.

Figure 3.1 is not complicated by separate boxes for the milk, eggs, and semen that are vital to satisfactory production of succeeding generations of animals.

3.1.8 Fertilizers

The inflows to the 'soils' box of Figure 3.1 include, as already noted, the *plant residues* returned to the sites where they grew. Some crops are grown for use entirely as 'green manure'. In addition, the *mineral nutrients* available to plants in soils are increased by those slowly released by chemical weathering of rocks and soil particles. Additions of *nitrogen* result from the washing down of nitrogen compounds from the air, and the chemical fixing of nitrogen itself by bacteria in the root nodules of plants of the legume family, or by other organisms in the soil. From the *wastes and excreta* box there are important fertilizers in the form of farmyard manure (dung), night-soil, kitchen wastes, and other materials such as meat and bone meals. And of great importance to modern agriculture are the products of *fertilizer industries*, containing the following: nitrogen fixed from air, mined as sodium nitrate or guano, or obtained in industrial by-products; phosphates, mined or obtained as a by-product of the steel industry; potash, mined, or obtained from the Dead Sea; calcium mined as limestones; other elements in smaller quantities but important for some areas. All these are inflows that serve to correct chemical deficiencies in soils for crop plants. In addition physical and biological deficiencies can be corrected by use of organic materials prepared by the fertilizer industries, including some of the wastes and excreta already noted, and horticultural peat. If soils appear to lack the *Rhizobium* bacteria that fix nitrogen in legume roots, these bacteria are added with the legume seeds.

3.1.9 Rocks

We have already noted the importance of rocks as the parent materials of soils, as slowly continuing sources of plant nutrients, as affecting underground water flows and reserves, and as the sources of phosphates and other raw materials for the fertilizer industry. Rocks are important in many areas also as providing building materials affecting the biological sub-system through the work sub-system (e.g. when used for building field walls, terraces, and water collection systems). Stones are important in reflecting and radiating heat to help ripen grapes in some high quality wine-producing areas. Elsewhere they may be obstacles to full mechanization of farming operations (e.g. potato harvesting).

3.1.10 Weeds, insect and other pests, diseases

Almost all the interrelationships shown in Figure 3.1 are affected by organisms that compete with the plants and animals of agriculture, parasitize or eat them, destroy agricultural resources, or damage agricultural outputs. The number of species is great and they vary widely in life histories, methods of reproduction and distribution, the physical environments and food supplies that favour them, and natural population controls – from those of organisms smaller than viruses to those of wild elephants. Some affect agriculture not only directly but also through their effects in humans (e.g. tuberculosis, brucellosis, ergot).

Many decisions about plant and animal production are therefore based on the need to avoid or control *the biotic factors*. Many decisions about the storage, processing, and transport of agricultural products have also to be based on the need to control these factors.

Some diseases are not due to organisms but to chemical inadequacies or environmental stresses, or inherited defects (Sections 4.2, 4.9.7, and 5.6).

3.1.11 Hydroponics

The water, nutrients, oxygen, and root-holds that plants derive from the soil are provided for some intensively produced crop plants by water solutions of controlled nutrient content, passing along narrow, well-aerated channels. The nutrition of plants such as glasshouse tomatoes can thus be well controlled, and soil-borne pests and diseases avoided. Where soil resources are extremely scarce relative to the human population and their purchasing power, green vegetable production by hydroponics is practised (e.g. at some Pacific naval bases).

3.1.12   Fish farming

This can be closely integrated with crop production (e.g. in the Yangste valley where small ponds receive various 'wastes' and produce carp, as well as mud used to fertilize the seedbeds of rice). Alternatively, fish production can be based on water resources not otherwise used for agriculture, and feedingstuffs derived largely outside agriculture such as sea-fish by-products (e.g. salmon production in the sea in net cages).

3.2   INTERDEPENDENCIES

3.2.1   Within the biological sub-system

Our brief review of the biological sub-system has indicated many dependencies. Humans depend on plants and animals. Animals depend on plants. Plants depend on climates and soils. Soils need replenishments in various forms. Weeds, pests, and diseases arise throughout the system. Pollutions and depletions can occur.

If we consider the *cycles for all the individual chemical elements*, and those for energy and water, the interdependencies seem even greater. (For example, the nitrogen (N) cycle and the potash (K) cycle both provide essential supplies to crop plants, and suitable ratios of K to N are essential to plant health. But crops differ in N and K contents, and so do the replenishing flows into soils from various sources.)

Other reasons for complex dependencies are the *seasonal changes* in climates making, for example, animals dependent on grazings in summer and stored feedingstuffs in winter, while human diets also change substantially with the seasons. *Between years* there are obvious dependencies (e.g. the long-run effects of improving or depleting soils, controlling pests, or keeping more livestock).

And interdependencies *between particular localities, areas, and regions* result from trading.

3.2.2   With other sub-systems

Our brief review has also indicated close connections between the biological sub-system and the *other sub-systems* of agriculture. Apart from natural interconnections of water supplies to climates, all the lines in Figure 3.1 are affected by the work sub-system. For example, the relations of plants to climates are affected by all the tillage, fertilizer distribution, sowing, and other operations that aim to present, at the right times, the right cultivars, at the right stages of growth and development, so as to make full use of climatic and soil resources. Much work and equipment is required to secure feedingstuffs for animals and to house, feed, and manage them. The farm-economic sub-system is affected by all the biological flows because they arise from natural endowments and flow resources that are scarce. The outputs produced therefore have values. Decisions in the farm-economic sub-system are about these values and affect priorities in the biological and work sub-systems (e.g. decisions about what to produce to provide the largest possible farm family incomes during a particular period in a particular locality). And many problems in the socioeconomic sub-system (e.g. about markets, land tenure, research) are intimately connected to conditions and relationships in the biological sub-system as well as those in the work and farm-economic sub-systems.

3.3 ILLUSTRATIONS FROM EXAMPLE FARM GROUPS

From the nine widely differing areas that were briefly described in Table 2.1, the example farms illustrate important characteristics of the biological sub-system. Tables 3.1–3.5 summarize data on land use, cropping, animal species and numbers, biological inputs, and outputs and their composition.

The wide differences in climates and soils are largely matched by a complex pattern of the crop species for which the climates and soils are used. Even where

Table 3.1   Land use (nine example farm groups, 1978–9).

	Tilled crops	Permanent crops	Rotational grass [a]	Permanent grass	Rough grazings	Total area[b]
			per 100 ha of total area[b]			*hectares per farm*
Aberdeen	38	0	51	11	27	97
Central Norfolk	88	1	3	8	+	122
Cortland, NY	40	0	35	25	17	164
Central Illinois	97[c]	0	1	2	0	166
Saskatchewan	98[d]	0	+	2	2	345
South-western Louisiana	82	0	0	18	14[e]	295
Iloilo	100	+	0	0	[f]	3
Aurepalle	84	0	0	16	+	5
Western Malaysia	1	99	0	0	+	3

[a]Including alfalfa.
[b]Excluding rough grazings, woods and forests.
[c]Including 1 ha under diversion agreements.
[d]Including 37 fallowed.
[e]Including fallow and 'idle' land.
[f]'Common grazings' not included.

Table 3.2 Crop species on tilled land (nine example farm groups, 1978–9).

Areas	Wheat (w), rice (r)	Barley	Other small grains	Maize (m), sorghum and millets (s)	Other crops Main	Miscellaneous
			hectares per 100 ha of tilled land			
Aberdeen	2 w	78	7	0	12 turnips	1
Central Norfolk	30 w	35	1	0	{ 3 potatoes, 18 sugar beet }	13
Cortland, NY	2 w	0	9	{ 41 m silage, 47 m grain }	0	1
Central Illinois	+ w	+	0	52 m	44 soya	3
Saskatchewan[a]	46 w	6	1	0	{ 4 rape, 4 flax seed }	1
Louisiana	33 r	0	0	+	67 soya	+
Iloilo[b]	125 r	0	0	68 maize, beans, yams, etc.		
Aurepalle[c]	20 r	0	0	24 s	33 castor	12
Western Malaysia	30 r	0	0	0	0	70

[a] Not including 38 fallowed.
[b] Total area of crops was 193 per cent of tilled area, due to multiple cropping.
[c] 19 per cent of tilled area was fallowed; 5 per cent, double cropped; 30 per cent, inter- or mix-cropped.

Table 3.3 Animal species and numbers (nine example farm groups, 1978–9).

Areas	Cattle[a] Cows	Other	Sheep, goats	Pigs	Poultry	Total animal units[b]
	numbers per 100 ha of total area					*number per farm*
Aberdeen	41(B)	142	179	7	32	148
Central Norfolk	13(D)	18	+	18	+	60
Cortland, NY	56(B)	23	0	0	0	130
Central Illinois	←Very few animals→					+
Saskatchewan	←Very few animals→					+
Louisiana	←Very few animals→					+
Iloilo	0	48(C)	←Not enumerated→			1+
Aurepalle	14	92[c]	63	0	42	5
Western Malaysia	0	0	Not enumerated		228	+

[a] (B), beef cows; (D), dairy cows; (C), water buffalo for work.
[b] Approximate. The units per animal have been counted as follows: dairy cow, 1.0; beef cow, 0.8; buffalo, 0.8; other cattle, 0.6; sheep or goat, 0.18; pig, 0.15; poultry bird, 0.01.
[c] Including 33 bullocks, 47 buffalo, and 12 young cattle.

Table 3.4 Biological inputs (nine example farm groups, 1978–9).

Areas	Total per ha of total area	Ferti-lizers[a]	Seeds	Sundry, for crops	Feeds bought	Veter-inary Services[b]	Sundry for animals	Total
	US$			US$ per US$100 gross output				
Aberdeen	246	10	4	2	9	1	3	29
Central Norfolk	369	7	5	7	7	1	1	28
Cortland, NY	406	5	2	2	21	2	5	37
Central Illinois	132	9	←8→			←Very little→		18
Saskatchewan	20	6	3	5		←Very little→		14
Louisiana	96	3	8	11		←Very little→		22
Iloilo	26	1	5	+	+	0	+	7
Aurepalle	25	5	3	1	8	+	0	18
Western Malaysia	29	←6→			0	0	0	6

[a]Including lime.
[b]Including drugs, antibiotics, etc.

Table 3.5 Quantity and composition of gross output (nine example farm groups, 1978–9).

Areas	Total value per hectare of total area	Wheat (w), rice (r)	Maize (m), other grains	Other crops	Milk cattle, hides	Sheep, wool	Other animal products	Total value per farm[b]
	US$		US$ per US$100 gross output					US$'000
Aberdeen	832	...	14[a]	2	64	10	2	81
Central Norfolk	1324	16w	9	37	16	+	19	161
Cortland, NY	1099	1w	3	3	92	0	0	180
Central Illinois	731	+w	51m	47	+	+	+	121
Saskatchewan	146	62w	11	9	+	+	+	50[c]
Louisiana	447	49r	+	51	+	+	+	132
Iloilo	384	75r	12	13	←n.a.→			1
Aurepalle	141	30r	12	33	25	+	+	1
Western Malaysia	436[e]	1r	0	83[d]	←n.a.→		2	1[e]

[a]Including wheat.
[b]Including miscellaneous receipts.
[c]Including income from insurance companies, contract work, and other sources.
[d]Almost all rubber.
[e]Including off-farm work.

the tilled land is used for grain crops, the species are in widely differing ratios (Table 3.2). Thus most of the grain area in the Aberdeen group is barley; most in Saskatchewan, wheat. In Aurepalle, rice can be grown only if irrigated: sorghum and millet are the grain species on unirrigated land.

The grazings and other crops for animals differ in relation to crops for direct human use, and in feeding value. Therefore the use of animals and the relative importances of the different species are widely varied (Tables 3.2 and 3.3). We should note, of course, that some example farm groups are selling grain that is used as feedingstuffs by other groups, largely in other areas, and even in other countries.

What inorganic fertilizers, seeds, concentrated feedingstuffs, and other 'biological' inputs are brought into the production systems of different example groups also vary widely per hectare of land, and per US$100 of output of agricultural products (Table 3.4).

The main results of all the variations in crop and animal species used, and in inputs, are wide variations in the composition of the outputs and the total outputs per hectare (Table 3.5) – the total outputs increasing from US$141 per hectare in Aurepalle to US$1099 in Cortland, NY.

QUESTIONS AND EXERCISES

1. From a biological standpoint, what are the main outputs of the agriculture you know best, and what are the main determinants of trends and variations in these outputs? Explain briefly.
2. Contrast two sites or areas that are used for agricultural production. What are the main biological differences?

Chapter 4

Crop production and protection

4.1 DETERMINANTS OF BIOLOGICAL SUCCESS

The brief survey in Chapter 3 indicates that crop production will be *biologically* successful if the following principles are adhered to:

(a)   (i) Good use is made of the wide variety of natural endowments of climates and soils; and (ii) climates and soils are well improved for crop plants, full use being made of 'wastes and excreta' and crop residues
(b)   (i) Decisions are well made about which species and which cultivars to grow on which sites; (ii) improved cultivars are bred and selected, suiting particular sites better; (iii) decisions are well made about the spacing of individual plants, the use of mixtures, and the sequence of crops, because these all affect the climatic and soil conditions that individual crop plants experience
(c)   Biotic factors are well controlled
(d)   The timing of the life cycles of crop plants is appropriate to climates and soils not only because of (b), but also because tillage, sowing, and other operations are well timed
(e)   Basic endowments of climates, soils, and water are not depleted, nor atmospheres, soils, and water polluted

This chapter considers first (b), because we should have a general understanding of differences between crop plants, and ideas about spacing, mixtures, and sequences, before considering (a). We can then usefully consider (c), (d), and (e).

4.2 INHERENT DIFFERENCES IN PLANTS THAT AFFECT THEIR YIELDS

4.2.1 Phenological controls

The controls of biochemical and physiological processes in crop plants are sensitive and intricate, affecting the timings, rates, and extents of developments (e.g. in wheat plants the development of tillers (side-shoots); the initiation of flowering; ear structure; grain ripening). What differ are the responses of plants to environments. What are inherited are the biochemical controls of these responses. To illustrate the practical importance of these differences some examples are now given for wheat.

4.2.2 Eleven sets of differences in wheat

(a) *Timing* – of germination; start of vegetative growth after dormancy; tillering; leaf area development; start of reproductive processes; flowering; pollinations; start and finish of the storage (grain-filling) period, including ripening; senescence of leaves and of the whole plant. In regions about 30°N and 30°S, wheat has to develop vegetatively during short, cool, winter days of low light intensity. Flower development has to start as risks of frost are passed, days lengthen, and light and temperatures increase. Otherwise later high temperatures and lack of water would stop grain filling too soon. About 50–55°N in continental climates (e.g. in our Saskatchewan area) winters are long and severe. Wheats have to be spring sown and complete development and ripening within a short growing season. They start flower development in long days and have to fill the grains and ripen quickly. They too may be subject to shortage of water and high temperatures during the grain-filling period. Further south in the USA, in and around Kansas, winters are less severe and winter hardy varieties can be sown in mid-September and ripen by mid-June. In the maritime climates of western Europe the growing season can be still longer (e.g. in our Norfolk area). Wheats are autumn sown and grain filling can continue until well after mid-summer under cool conditions. The temperatures required for ripening are lower than those for Red Winter wheat cultivars in Kansas.

(b) *Structure and size* – depth and 'density' of the root system; tillering; leaf area, form and inclination; branching; height and structure of the canopy (crop cover); relative numbers and storage capacities of the parts that are 'output'. For many areas the improved wheat varieties have shorter straw and a lower straw : grain ratio. They tiller less than older varieties, and have a smaller root system. They invest therefore less of their assimilates in vegetative growth and relatively more in grain output. Their yield of grain is higher per hectare, and higher in relation to the water used. The leaves of improved varieties are held more nearly vertical so that light is more uniformly distributed through the canopy, and photosynthesis and yields are higher. Some breeding programmes now aim to increase total plant size again.

(c) *Net rates of assimilation* – synthesis of carbohydrates and other compounds – summation of short-term durations and rates in response to environmental conditions, less losses through respiration. There are major differences between (i) maize, sugarcane, and other tropical grasses and (ii) almost all other crop plants. Group (i) are called C_4 plants because in an intermediate molecule in the photosynthetic process there are four carbon atoms. Group (ii) are called C_3 plants. In the high solar radiation and temperatures of the Tropics, the C_4 plants have higher photosynthetic rates and lower respiration rates in daylight. C_3 plants may be less productive at high temperatures. But at lower temperatures, and with lower light intensities, they are more favourably placed. Wheat is a C_3 plant and it has not yet been possible to raise the short term photosynthesis per unit area of flat leaf. But yields have been raised by selecting varieties with more nearly vertical leaves and longer grain-filling periods.

(d) *Reactions to physical stresses* – particularly to water shortages, and flooding; low and high temperatures; wind; hail and snow.

Winter-hardy wheat varieties have been secured for various important regions, and short-strawed varieties are grown where water supplies are liable to be low. Stiffness of straw as well as shortness helps to reduce damage to grain and losses due to wind and rain. Some breeding programmes aim at greater ability to extract water from drier soils.

(e) *Reactions to biotic stresses* – shading and other competition from plants; pests and diseases; heavy grazing by animals; frequent cutting, pruning.

The biotic stresses on wheat crops are reduced by agronomic practices including tillage, rotation of crops, use of clean seed, and herbicides. Resistance to fungal diseases (e.g. rusts) and to Hessian fly attacks are bred into many modern wheat varieties. Some improvement programmes aim at mixed types of resistance to rusts because the resistances in a completely uniform cultivar can fail when the parasitic organisms themselves mutate so that the detailed chemistry of their attack changes. There are more than 300 known races of stem rust in wheat, but they are not all present in any particular region at any one time.

(f) *Sharing out of assimilates* as between different parts of the plant, including those that are 'output'.

Again the controls that result in low straw : grain ratios are important. The location of photosynthesis near to the wheat ears tends to increase grain yield.

(g) *Rates of transport of assimilates* are increasingly studied. They appear often to limit grain filling in wheat crops.

(h) *Structures, form, and chemical composition of the 'output'*. Wheat varieties differ in the number of heads (ears) per plant and the number of grain-bearing florets per head. They differ also in how much grain is shed too soon from the ears. One of man's first selections was in effect against wild types that shed their grain easily. Modern wheats also differ in starch and protein contents of the grain and in the quality of the protein as shown when the respective flours are used (Section 3.1.2). 'Hard' red wheats are especially good for bread; 'soft' wheats for pastry and biscuits; Durum wheats for spaghetti and other pastas. In other crops it may be essential to breed out bitter or poisonous contents.

(i) *Other structural characteristics as affecting harvesting*. Many crops are liable to insect and fungal attacks that reduce the qualities and quantities of output. Qualities may also be lowered by rain or snow if harvesting is delayed. Therefore not only should crops ripen in good time, but their structures, and stances, should make them readily harvested by whatever methods are intended.

Again, short stiff wheat straws with ears of grain held erect and grains not shedding too early are important, especially where harvest periods are liable to be windy and wet at times, or much restricted by the early onset of winter.

(j) *Adaptability* – to wide differences in environments, and therefore *reliability* of yields and qualities. The adaptabilities of the main crop species are much greater than those of the wild species from which they were derived. This is because (i) different 'ecotypes' have been selected and bred for particular environments, and (ii) some other 'ecotypes' have been selected and bred for wider adaptability. Thus, in Mexico, wheat breeding programmes aim to secure that the timing of the start to flower development and the pace of development are independent of day length.

(k) *Persistence.* Long life in 'permanent' crop plants is desirable – for example in grasses in pastures; fruit trees; rubber trees; tea, coffee, and cocoa bushes. In annual or bi-annual crops good seed setting is desirable.

4.3   GENETIC BASIS OF CROP PLANT IMPROVEMENTS

Plant breeders have substantial opportunities to select from wild species and from cultivars those characteristics that will constitute their ideal when well assembled in one plant or population. There is enough variation to permit this (e.g. in maize, more than 8000 varieties).

A basic understanding of the opportunities and the problems in using them requires an introduction to the chemical basis of heredity. *Genes* in individual cells of plants are particular sequences in molecular chains of DNA (deoxyribonucleic acid). These chains of DNA are complexed with proteins known as 'histones' to form *chromosomes*. Individual genes control specific functions because they are replicated when cells divide and plants grown, and because they specify which proteins are synthesized (including which enzymes) and when. Enzymes control the timing and form of developments.

Chromosomes are usually found in pairs. Maize has, for example, 10 pairs of chromosomes, and over 500 identified genes. The individual chromosomes of each pair may both have a particular gene in one or two or more alternative forms (alleles). So a plant may be *homozygous* or *heterozygous* for a particular allele of a particular gene. In sexual reproduction an interchange of alleles may take place because one member of each pair is separated from the sperm cells in the pollen, and one for the eggs in the ovaries. A fertilized egg may therefore have particular allele pairs that are different from those in the parent plants. Thus, to explain the results that Gregor Mendel obtained in crossing (hybridizing) tall and dwarf peas, we can represent the gene controlling height as TT in homozygous tall peas or as tt in homozygous dwarf peas. The first hybrid generation (F_a) all had the Tt pair and were tall because the T allele is dominant in its effects. But the second generation (F_2) obtained by controlled self-fertilization produced three different combinations because the Tt pairs, after separating into single genes in single chromosomes, recombined into TT, Tt, and tt pairs. These were in the ratio $1:2:1$ because chance determined the recombinations:

		Pollen cell chromosomes	
		T	*t*

		Fertilized egg chromosomes	
Egg cell	*T*	*TT*	*Tt*
chromosomes	*t*	*Tt*	*tt*

In the third generation (F_3), obtained by self-pollination of F_2 individuals, the *TT* and *tt* individuals 'bred true' because they were homozygous. The *Tt* individuals being heterozygous produced the same $1:2:1$ ratio as the F_1 individuals did.

Occasionally sudden changes (*mutations*) occur. If the change is within a gene it is called a *point mutation*. But the structures of chromosomes may also change, so that sections disappear, or are repeated, or their order along a chromosome is altered, or sections may switch between different chromosome pairs. Also, instead of the usual two sets of individual chromosomes (two *genomes*), mutations or interspecies crosses may produce three or more sets (*polyploidy* instead of *diploidy*). Our main wheat species has 42 chromosomes made up of three sets (genomes) of chromosomes each of seven chromosome pairs. The three *genomes* came originally from three wild species each of which had seven chromosome pairs. Some interspecies crossing is possible.

Some progress may also be secured by inducing mutations by physical and chemical means and by developing techniques to transfer genes via the chromosomes of bacteria. But so far there is little control of the results of these methods.

4.4   CROP PLANT IMPROVEMENT PROGRAMMES

4.4.1   Method of reproduction

Much depends on whether a crop is naturally cross-pollinated or self-pollinated, or can be commercially propagated vegetatively. Potatoes are vegetatively propagated by planting tubers. Thus once an improved variety (cultivar) is selected – after sexual methods of hybridizing – it can be mutliplied up without need of further steps to control the genotype. Wheat is a largely self-pollinated crop. In a breeding programme desirable selections from the F_2 generation may well not be homozygous. For some five generations further selections are necessary so as to discard plants with undesirable genes. Maize is cross-pollinated so that the development of true-breeding cultivars requires still more care. Pollination has to be artificially controlled for about five generations and reselections made in each generation. Vigour and yield tend to decline as homozygosity is approached. Moreover, even after the five generations, care to avoid pollination from other cultivars continues to be necessary to maintain good seed stocks.

4.4.2 Use of hybrid vigour

Fortunately, use can be made of the hybrid vigour that results from heterozygosity in maize and some other cross-pollinated crops. Selected plants are inbred for several generations to establish selected inbred lines. The selection procedures aim to secure lines that will on crossing complement each other to provide substantially improved hybrids. Double crosses (hybrids of two hybrids, each from two inbred lines) and other uses of inbred lines are made. Between 1935 and 1970 the percentage of the maize area in the USA that was sown to hybrid maize increased from 2 to 99 per cent. And since 1955 in other countries much progress has been made in hybrid maize production.

Use cannot be made of hybrid vigour in self-pollinated crops unless effective and sufficiently cheap ways can be found to render some of the inbred lines self-sterile. Progress is being made for wheat, rice, and some other important crops.

4.4.3 Some basic problems

The work of crop plant improvement is in practice inevitably complex and time consuming, whether or not the aim is to produce commercial F_1 hybrids.

Desirable characters are often not controlled by single genes. The results of selection and hybridization may be quantitative and difficult to measure reliably, rather than obvious. The desirable genes may occur in existing plants in awkward combinations with undesirable genes. To secure the desirable combinations very large numbers of plants have to be grown and carefully observed. Rigorously designed experiments are commonly necessary to secure reliable yield comparisons. There may also have to be chemical testing for qualities. The ideal crop plants may themselves be difficult to define well, because there is not yet adequate understanding of the determinants of yields and qualities. Moreover, the definitions may have to be changed because the biological environments that they are to suit are changed by improvements. Mutations in fungal and other disease organisms may make previous conceptions of the desirable gene complements for resistance to them seriously inadequate. Changes in possibilities of mechanical sowing and harvesting may also alter definitions. Also changes in market preferences may alter the premia for particular qualities in yield. And not least are the problems in deciding how wide is the range of climatic, soil, and other conditions that cultivars should be bred to suit. Adaptability is a useful aim, but the best possible yield in any particular area cannot be obtained without breeding and selection to suit its conditions.

4.5 SPACING

To make full use of climate and soil, crop plants have to be correctly spaced. If too widely spaced their total leaf areas per hectare are insufficient to make full use of light, especially early in the growing period, and weed growth may be increased. If too closely spaced, competition between plants for light, water, and nutrients may reduce the total yield of the output parts and lower quality. Fungal diseases may be favoured by the denser vegetative growth and higher humidity in the canopy. Where water supplies are limiting wider spacing is desirable. The best spacings differ between species and between cultivars of the same species, and between regions. In grape, apple, and other orchard production, correct spacings (including spatial arrangement) are important as contributing, along with correct pruning, to full ripening of fruit.

Spacing practices are influenced by conditions in the work sub-system. For example, where weed control is by hand labour, a balance has to be struck between (i) the possible increases in yield from more plants per hectare and (ii) the extra work of weeding round more plants and the difficulties of ensuring, in the awkwardly restricted spaces between them, that weeds are killed by drying. Where cultivations and harvesting are mechanized, the balance may be between (i) the biologically most effective spacing, (ii) the spacing best suited to the available machines, and (iii) the re-design of machines.

4.6 MIXED CROPPING

Plants of the same cultivar compete with each other more directly than do plants that differ in canopy structure, timing of growth and nutrient requirements, depth of rooting, etc. Therefore over 12 months on one particular site, fuller use may be made of the total available light, water and soil nutrients if two or more species or cultivars are grown together.

The most common type of mixed cropping in temperate regions is in rotational grassland (ley) production (Table 3.1). Several species of grasses (with perhaps two cultivars of one or two of them) and one or more species of legume are commonly sown in spring with small grains (e.g. barley). Total production is substantially higher than it would be from single cultivar stands.

In most tropical countries mixed cropping is common (Table 3.2). Many climatic conditions cannot be fully used by single *annual* crops because their leaf area is insufficient for too long. Also ground cover at all times is desirable to reduce soil erosion, and solar radiation to the soil, and to suppress weeds. Mixed cropping also slows up the spread of pests and diseases, because different crop species have different susceptibilities. Also, if one species fails, others can make use of the 'plant space' left. Where *perennial* bush and tree crops (e.g. cocoa, rubber) and long duration crops (e.g. cassava) are grown, annual crops are commonly mixed with them so that full use can be made of the flow resources during the early years. Many clever, complex cropping systems have been evolved to suit the conditions of particular areas (e.g. in our Iloilo and Aurepalle areas).

Mixed cropping does, however, have biological disadvantages in that (i) pesticides, fungicides, and fertilizers have to be used largely for the 'mix' and not more precisely for the individual species; (ii) mixtures may suffer greater stresses due to drought than selected individual species; (iii) agronomic research and development, and plant breeding work, are made especially difficult (e.g. grassland research).

Also many mixed cropping plans make mechanical planting, weeding, and harvesting impossible. But operations in mixtures have to be carefully timed. On the other hand, mixed cropping can reduce the peaks of labour requirements (e.g. for early weeding) that are serious in many single crop plans.

4.7 CROP SEQUENCES

4.7.1 Multiple cropping

Another way of making fuller use of flow resources is to grow crops in quick succession so that more than one harvest is secured within 12 months. Thus in the Tropics, two or even three crops of rice may be obtained each year if cultivars with suitable phenological controls are used and all nursery bed and field operations are well timed (e.g. in Iloilo). Five rice crops in three years are now common in Java. Winter crops of wheat and summer crops of rice are common on irrigated sites in northern India, Pakistan, and central China.

In temperate regions multiple cropping is common in horticulture, and particularly under glass, with careful choice of species and cultivars, and careful timing.

But under temperate field conditions the usual aim for each site is only one crop to harvest each year, with careful decisions about the succession of crop species over the years.

4.7.2 Crop rotations

When the succession is repeated regularly there is said to be *a crop rotation*. In many temperate areas rotations are less rigidly practised than in the past, when there was little or no use of inorganic fertilizers and powerful equipment for tillage, and no herbicides and pesticides. Even with these aids, the advantages of rotations are still widely sought. And in the Tropics many regions with rising human populations have to try to secure them. A rotation that did much to improve English agriculture was adopted about 1730, the Norfolk four-course rotation – Wheat, Turnips, Barley, Clover with sown grasses. The biological advantages were essentially in maintaining soil fertility at a higher level and making good use of it, in integrating animal and crop production; in feeding animals much better in winter; in controlling weeds, and crop pests and diseases; and in securing better timing of field operations.

Similar basic advantages are obtainable in many regions. Special attention may be given in particular areas to reducing water or wind erosion of soils, keeping up the organic matter in soils, conserving soil water supplies, and avoiding high weed pest and disease populations. Thus in central Indiana, USA (in a rotation of maize; maize; soy beans; wheat; hay) cover crops are sown in both the maize crops and phosphatic fertilizers as well as farmyard manure are applied. In the south of the USA soil conservation and soil fertility are objectives in sowing winter cover crops of legumes and rye, and ploughing them in as green manure. In the Great Plains water is conserved in the drier areas by the wheat; fallow rotation. In areas with more precipitation, the rotation can be wheat; sorghum; fallow. With still more it can be wheat; maize; fallow.

4.7.3 Shifting cultivation

This is commonly practised in the Tropics. Patches of forest or savannah bush are cleared by cutting and firing. After about 3–5 years of cropping the patches are allowed 'to rest'. Wild species cover the soil slowly in the drier savannah regions but faster in the forests. Soil nutrients and some organic matter build up in the soil again. The lengths of the cropping and 'rest' periods depend on densities of human population on the land, and land tenure arrangements. If the 'rest' period is relatively long, soil fertility may be restored; but if human populations increase and the rest period is progressively shortened, soil fertility will decline unless fertilizers are available and used. Commonly fertility has declined, lowering crop yields and reducing the growth of wild species. Thus the effects of 'rest' periods are reduced. And commonly weed control becomes more difficult because the species that do grow are less easy to remove by fire or hoe.

The sequence of crops has important effects in the work, farm-economic, and socioeconomic sub-systems. The spreading of labour requirements more evenly during a year has already been noted. The increase in total outputs and diversification both have important effects in income levels, risks and uncertainties, and conditions for trade.

4.8   USES AND IMPROVEMENTS OF CLIMATES

The natural determinants of climates are so powerful and widespread that mankind has not yet been able to improve climates much but has rather had to learn to use them. Yet some important improvements have been made and will be extended, provided that they are judged sound, on engineering and economic reasoning as well as biological.

4.8.1   Water

Irrigation, flood control, and *drainage* are major improvements related to climatic conditions but best considered along with improvements to soils (Section 4.9.3). 'Rain making' by the 'seeding' of clouds has narrow natural limits and high costs. Some advances have been made in preventing *hail* storms that can do great damage to grapes and other fruit. Evapotranspiration from crops is reduced by shelter belts, shade trees, and other forms of shelter and shading, and evaporation from the soil by *mulches* (e.g. of polythene, crop residues, bark chips).

4.8.2 Temperature

Shading, *shelter*, and *mulches* also affect temperatures.

Glasshouses allow radiant heat from the Sun to be absorbed during day time, but do not allow long-wave radiations to pass heat back from inside at night. Glasshouses can also be used to shade and cool plants when outside temperatures are too high. Cold frames, cloches, polythene houses and tunnels, and 'hot caps' are important uses of the same principles for high value crops. The soils in glasshouses are *heated directly* when propagating some crops in Europe. Even air heating is judged economic in periods critical for flowering in some orchards and in some vineyards when young shoots appear in spring. When heat is critically short for valuable crops, *irrigation* water may also be used to bring it to a site. Irrigation may also improve the conduction of heat upwards within soils and subsoils.

4.8.3 Light

Complete or partial shading is used in horticulture (e.g. to blanch endive and chicory) and in tobacco production to secure high quality cigar 'wrappers'.

4.8.4 Light and temperatures

When close controls of flowering times are required, light or temperature or both may be controlled in detail by shading or supplementary lighting, and by heating or cooling. The inbred phenological controls of the plants, if appropriately stimulated in these ways, result in valuable responses (e.g. in cut flower production out of season; in providing pollen and ovaries when required for hybridization work). Species that start to flower when days are short or shortening (e.g. Chrysanthemums) can have the natural nights lengthened by shading. Long-day species can have supplementary lighting.

4.8.5 Wind

Shelter belts and screens are sometimes used to reduce direct physical damage to crop plants as well as to reduce evapotranspiration and modify temperatures, and wind erosion of soils.

4.9 USES AND IMPROVEMENTS OF SOILS

4.9.1 Soil conditions and plant growth

We have already noted that the natural determinants of soil formation have produced an intricate pattern of soils over the Earth's surface (Section 3.1.5). Mankind has been able to improve soil conditions substantially and successfully.

Agriculture depends on understanding the possibilities of improvements. In this chapter only biological aspects are considered.

Plants obtain the following from soils: water; oxygen; major nutrients; minor nutrients; root holds for security; and sites on which to develop their above-ground canopy. The provision of all of these should be favoured and not obstructed by the conditions in soils.

Therefore improvements (i) make good deficiencies (e.g. of water or phosphate); (ii) make some of the physical, chemical, and biological conditions better (e.g. temperature, drainage, bacterial activities); (iii) remove some obstructions (e.g. excess acidity or salt content, disease organisms, weeds). Each practical improvement commonly affects crop growth in more than one way, because soil conditions and what plants obtain are usually all closely interconnected (e.g. drainage improvements affect soil temperatures in spring, oxygen supplies, acidity, and releases of nutrients). The various needs of plants and soil conditions can therefore usefully be reviewed together, and with the main practical means to improvement. Table 4.1 provides a matrix for this purpose. The number of crosses in the 'pockets' gives some indication of which fertility determinant is most affected by each practice. The improvement practices that are most important in Europe and North America are asterisked. Some especially important in the Tropics are noted below.

4.9.2 Temperatures

Temperature affects germination, root development, and root functions, and also chemical and microbiological actions in the soil. Thus for wheat germination, the minimum temperature is not much above 0°C but the optimum is much higher. For maize and sorghum the minimum is about 9°C. The breakdown of crop residues and dung by soil organisms is more rapid at high temperatures.

Soil temperatures may therefore be improved in temperate regions by raising them. Direct heating and the glasshouse principle are used for high value horticultural crops. Flood control, drainage, and the making of ridges can raise temperatures by reducing the water in soils, because more heat is required to raise the temperature of water than the temperature of soil particles and the air between them. Tillage can also help by increasing drainage and aeration and so the drying of the upper soil.

In the Tropics improvements by lowering soil temperatures are common: mulching; shading by cover crops, shade trees, and nursery screens; mixed cropping.

4.9.3 Water

Useful distinctions can be made between the following: (i) water so closely held in 'dry' soil by soil particles and organic matter that it cannot be taken up by crop plant roots; (ii) water vapour in the soil atmosphere; (iii) water that can be taken

Table 4.1 Soil management practices to sustain or improve soil fertility.

Practices	Temperature	Water	Depth	Structure	Nutrients	pH
				Fertility determinants		
Flood control	× ×	× × ×	× ×	×	×	×
Drainage*	× ×	× × ×	× ×	×	×	×
Tillage*	×	× ×	×	× × ×	×	
Sub-soiling		× ×	× × ×			
Deep ploughing		× ×	× × ×			
Irrigation*	×	× × ×				
Glass, polythene*	× × ×					
Heating	× × ×					
Ridging	× ×	× ×	×		×	
Mulching	× × ×	× × ×		× ×	×	
Shading	× × ×	× ×				
Crop residues*		×		× × ×	× × ×	
Dung, urine*		×		× × ×	× × ×	
Composts, night soil		×		× × ×	× × ×	
'Catch cropping'		×		× × ×	× × ×	
Avoiding poaching*		×		× ×		
Nitrogen, phosphorus, and potassium*				×	× × ×	
Minor and trace nutrients				×	× × ×	
Flooding for silts					× ×	× × ×
Lime*						× × ×

Practices	Toxins	Micro-biological	Other flora and fauna	Weeds	Erosion control[a]
Drainage*	× ×	× ×	× ×	×	× ×
Tillage*		× ×	× ×	× × ×	×
Crop residues*		× × ×	× ×		
Dung, urine*		× × ×	× ×		
Composts, night soil		× × ×	× ×	[a]	
'Catch cropping'		× × ×	× ×		
Nitrogen, phosphorus, and potassium		× × ×	×		
Minor and trace nutrients		× × ×			
Lime*	× × ×	× × ×	× ×	×	
Gypsum and flooding	× × ×				
Chemical and heat 'Sterilizations'		× × ×	× × ×	× ×	
Herbicides*				× × ×	
Rotations*		× ×	× × ×	× × ×	× ×
Wind and water barriers*					× × ×
Contour ploughing*					× × ×
Terracing					× × ×

[a]In general improvements in soil structure help to reduce both wind and water erosion.

up by roots, but held against gravity and therefore not draining (percolating) away; (iv) water that is draining away or would drain away if not stopped by water (or an impervious layer) lower in the soil; (v) run-off water that cannot enter (infiltrate) the soil surface because the total water in the soil is too great, or because the surface is not permeable enough to take fast enough the rainfall or other additions of water at the surface.

Improvements aim to ensure that supplies of type (iii) water are optimum for crop plants; type (iv) water is removed at optimum rates by drainage; type (v) water is minimized, and not allowed to cause soil erosion.

Direct improvements are noted first because, when they are needed, indirect improvements are difficult to secure.

4.9.4 Flood control, drainage, and irrigation

Flood control is essential to protect many soils of high potential productivity.

Rice has special physiological arrangements to supply oxygen from the air to its roots under water, and some rice cultivars can float. But execessive flood water can greatly damage other rice cultivars, and carry away plant nutrients.

Drainage improvements are needed in various circumstances. In some soils water percolates so slowly that their upper layers reach their capacities for water too quickly and run-offs are excessive. Some soils have a layer of iron compounds that prevents drainage. Some sites are excessively watered by seepages from higher up slopes. Where irrigation is practical there may be damaging accumulations of salts so that flooding and drainage are required to remove them. Excess water may also induce the build-up of other toxic conditions and favour some diseases.

Surface systems of raising beds of soil, or ditching, are used and also underground systems of tiled drains, or untiled (mole) drains. Pumping may be necessary (e.g. from the polders of the Netherlands). In the USA some 30 million hectares have been drained: in the United Kingdom over half the total arable land. In the old rice paddy irrigation systems of South-East Asia as much care was taken over drainage as over irrigation itself.

Irrigation is necessary where the crop plants to be grown would without it not have enough water of type (iii). Calculations of where and when this would be so include assessments of: (soil moistures at the start of a period) *plus* (precipitation during it) *less* (run off, percolation, evaporation, and transpiration). Usually irrigation is done when about 50–60 per cent of the possible total water of type (iii) has been depleted. Timely judgements and action are needed on each site, using local experience with perhaps some help from instruments to indicate water contents of soils. Supplies of water for irrigation are determined by the land areas from which water flows (the *watersheds*) – their sizes, climates, topographies, rocks, and soil and plant covers. Careful attention should therefore be paid to water suplies and relationships (hydrology) in watersheds, and especially to human uses of water because these can be excessive, and to plant covers because these are often depleted but can be improved by man. Water flow from one watershed to another has been achieved in a few areas by tunnelling through mountains (e.g. in New South Wales, Australia, and Sri Lanka). The desalting of sea water is undertaken, at high cost, in a few areas of the Middle East.

The distribution of irrigation water on the land may be by flooding comparatively large areas or small (e.g. the small rice plots along contours in Java); flows into furrows; spraying under pressure; or trickling to individual plants. In some circumstances there is underground channelling of water (sub-irrigation). All these ways have biological aspects. For example, trickle irrigation requires less water than other methods because evaporation losses are less, and humidity in crop canopies is not raised to levels that encourage fungal diseases. Difficulties in controlling weeds may be greater with furrow irrigation. The methods of conveying water to fields may also pose weed control problems – as well as human disease problems (e.g. bilharzia).

4.9.5 Tillage

'Hard pans' develop in some soils due to the movement downward of clay particles, or iron compounds, or the pressure of tractor wheels. Such 'pans' restrict water movements and root growth and have to be broken by tillage (e.g. deep ploughing, sub-soiling), if crops are to develop fully. In preparation of sites for sowing or planting, tillage operations should be wisely chosen and timed.

For seeds and seedlings, water and oxygen supplies depend on 'seedbeds'. Granules and crumbs in the topsoil should be small enough, yet not too loose nor liable to consolidate. The soil below them should be firm, but not compacted. The movement of water will then not be restricted and seeds and seedling roots will have enough water and enough oxygen. Roots can spread, and shoots emerge without having to force through a compacted soil 'cap'. Good 'seedbeds' depend commonly on other methods of soil improvement as well as careful and well-timed tillage.

4.9.6 Indirect methods to improve water relationships

These depend largely on the effect of colloidal matter on the *water holding capacity* of soils, and *soil structures*. The movement of water, oxygen, and other gases, and the penetrations of plant roots, all depend on the passageways between soil particles. Where these are too narrow or blocked, crop growth is restricted or fails. Some soils tend to have narrow passageways because of their *texture*, which is made up largely of fine particles (clays or silts) and because the clay particles are colloidal and swell when wet. Soils with coarse textures (e.g. sands) commonly have comparatively large spaces between their large particles. Passageways are usually satisfactory in loam soils because there are mixtures of particles of different size. Soils high in organic matter (e.g. muck and peat soils) have satisfactory passageways, provided that they are drained of excess water.

But much depends on *structure*. The widths and shapes of the passageways are determined by any aggregations of particles as well as by the sizes of individual particles. The clay colloids may be dispersed in water (deflocculated) and few particles of any kind aggregated. Or the clay particles may be held together (flocculated, or dried). Or they may be attached to larger soil particles, holding them together in aggregates. The fine residues of organic matter (humus) and gums and other organic substances exuded by bacteria, fungi, and plant roots help to stabilize such aggregates. Coarser plant or animal residues provide sources of humus, hold water reserves much as a sponge does, and keep passageways open. Numerous passageways are left by old plant roots and by fungi, worms, and other organisms. The soil particles that are not colloidal, and particularly the silts and sands, can be so compressed or tightly held by water of type (i) (see Section 4.9.3 above) that too few passageways remain sufficiently wide.

The ways to improve soil structures for crop growth are essentially ways of securing and maintaining the best sizes of aggregations of soil particles. They depend for success therefore on skilled and timely use of the physical properties of clay colloids, humus and other dead organic matter, plant roots, and soil organisms.

Thus in *soils with comparatively high clay contents*, the clay colloids deflocculate and aggregates break down when tractor wheels, ploughs, other tillage equipment, or cattle hooves press and 'churn' before the water content has been sufficiently drained and dried out. The passageways between soil aggregates become even more unsatisfactory if crop sequences do not provide penetrating root growth and crop residues enough, or dung. When wet (e.g. in winter in western Europe) such soils are sodden and sticky and if drying occurs in spring it is late, allowing only very short periods for tillage operations to secure seedbeds. Later tillage results in large clods of brick-like hardness. Improvements can be achieved by: ensuring adequate undersoil drainage, ploughing in early autumn when water content in and below the plough layer is not too high, and leaving a rough surface for freezing and thawing, wetting and drying; liming to deflocculate the clay colloids; avoiding crops like sugar beet which require much tillage and entail late autumn pressures from harvesting equipment; including crops that require few cultivations but provide much root growth and residues; applying dung; reducing wheel slip because this churns and smears the soil in the tracks; reducing the number of tracks.

Sandy soils often need improvement by a different set of measures. Their contents of colloids and organic matter are low. Methods of improvement have included, in Europe, large dressings of clay-rich materials (e.g. marl), and intensive and grass production for sheep, with crop residues and dung ploughed in. Modern, less costly methods, include: shallow autumn tillage; consolidation by minimum tillage for seedbeds and early seed sowing in spring; adequate fertilizer applications; rotational grass production for seed or livestock (where economic); and application of dung (where available). Wind erosion can also be reduced by leaving long straw stubbles anchored after autumn tillage.

Excess tillage. When aggregates are broken down more than is necessary for the plants being grown, increased damage to structure is probable from further breakdown in the topsoil, and unnecessary compaction from tractor wheels.

In the forest and savannah regions of the Tropics, axe, fire, and hoe are used rather than the plough. Some organic matter that could improve soil structure is lost by fire, but there is no disturbance of a plough layer of soil, nor compaction below it. Because of risks of water erosion and excess losses of soil organic matter due to high solar radiation, soils should be kept covered. Mixed cropping, hoe tillage, and use of cover crops and mulches therefore have substantial advantages.

In recent years in North America and western Europe interest in *minimal cultivation* and *direct sowing methods* has increased. These aim to reduce the breakdown of aggregates of soil particles by tractors and equipment. While weeds are controlled by herbicides or minimal tillage, roots and residues from above ground are used to improve structures and provide mulches. Success depends on careful judgements. Where herbicides or simple weed controls are effective, and other conditions favourable, the results can be satisfactory (and achieved at low costs). But where soil aggregates are liable to break down, drainage is poor, spring tillage has to be late, weeds are difficult to control, and the weather is wet – then the topsoil may become more compact and impermeable.

Rice padis pose some special problems. The best conditions include a surface layer of water with a population of algae that keeps the oxygen content of this water high. A layer of surface soil a few millimetres thick is also well oxygenated but below this is the main puddled soil in which water is retained and the first roots of the transplanted rice seedlings grow. This layer is short of oxygen but receives some from the rice roots which, unlike those of other crop plants, can draw on oxygen from air above the soil. Some oxygen also enters in water from the top layer, when bubbles rise from decaying green manures mixed with puddled soil. Below this puddled layer are layers of low permeability that allow slow drainage. After rice harvest the padi is allowed to dry out, securing fuller oxidation, and control of weeds. Poorer conditions do not provide enough water with enough oxygen, are not level enough for good water control, and do not allow enough drainage so that ferrous iron and other toxic compounds result. Careful judgements have therefore to be made about additions of organic matter, dates of flooding and puddling, dates of rice planting, uses of nitrogen fertilizers, drainage, running off water several times in the growing season, followed by reflooding.

4.9.7 Major and trace elements, pH, and microbes

Crop plants require many nutrients. *Major* elements are nitrogen, phosphorus, and potassium, which are often deficient in soils, calcium, magnesium, and sulphur, which sometimes are, and chlorine and sodium, which seldom are. Trace elements (with concentrations in plant dry matter of less than 100 p.p.m.) are boron, cobalt, copper, iron, manganese, molybdenum, and zinc (West, 1979). Attempts to make the nutrient availabilities (including the ratios amongst them) optimal site by site for the crops grown require many decisions. The best decisions depend on understandings and local measurements of soil chemistry, and its effects on plant growth. The following paragraphs state briefly the basic understandings required in our later considerations in this book.

Figure 4.1 elaborates Figure 3.1 a little to show systematically the major chemically interacting groups in soils and rocks. The soil *solution* can usefully be regarded as central, although the physical contacts between *roots, organisms, soil particles,* and *solution* are very intimate. Roots can be regarded as absorbing nutrients from the solution. The amounts and ratios in the solution are of key importance. These depend, as Figure 4.1 indicates, on (a) the chemical composition of the soil particles and rocks; (b) the adsorption, fixation, and solution (AFS) reactions between these and the solution; (c) the decay of organic matter (releasing nitrogen and mineral elements) brought about by living organisms – largely bacteria, fungi, and other micro-organisms (the M effects). This decay is also affected by the mixing activities of many other organisms, e.g. earthworms, termites, rodents; (d) the quantities and nutrient contents of organic matter for decay (including plant residues, dung, dead micro-organisms, etc.); (e) the activities of micro-organisms (NF) in fixing the nitrogen in soil air into ammonia, particularly the symbiotic *Rhizobium* bacteria in the root nodules of legumes (included also in (f)); (f) the symbiotic relationships (S) of micro-organisms, including fungi, in plant roots; (g) additions of inorganic fertilizers, including any used to affect soil acidity; (h) the water (with nitrogen, sulphur, and other compounds) entering as precipitation from the atmosphere; (i) the amount and distribution of soil air, with oxygen and carbon dioxide supplies; (j) the drainage away of the soil solution as against the upward movement of water due to evaporation, plant use and transpiration; (k) the uptake of nutrients by plants.

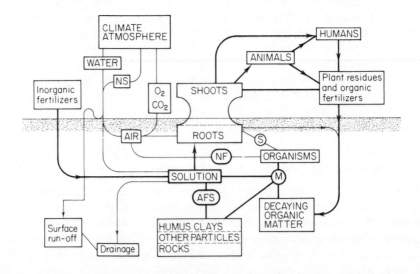

Figure 4.1   Major chemically interacting groups in soils.

The AFS, M, NF, and S reactions are all affected by (l) soil temperature; (m) soil water supplies and relationships; (n) the various ion concentrations in the soil solution in the recent past, including the hydrogen ion concentration (pH) which measures its acidity (pH 5 measuring strongly acid; pH 7, neutral; and pH 8, strongly alkaline).

This long list of the determinants of the fertility of soils from a chemical standpoint shows how closely related are the physical and biological determinants. It elaborates the main influences in soil formation (Section 3.1.5) and explains more fully why the natural variations in soil fertility are so great. Therefore it helps towards understanding maps of soils, from large scale maps such as farmers might use, to very small scale maps showing soil zones in the world (see Table 2.1, footnote *a*).

Excepting the natural endowments of soil particles and rocks, all the determinants (a)–(n) are *partly* under human control. *Improvements* are made by adding inorganic fertilizers (g), and by increased additions of organic fertilizers and crop residues (d), determined by decisions about species and cultivars, crop sequence and crop production practices affecting (k), the uses of crops, purchases of feeding stuffs grown elsewhere, animal species and numbers, and care of animal faeces and urine. The reactions at AFS, M, NF and S ((b), (c), (e), (f)) can be made more favourable by improvements to water supplies and relationships ((h), (i), (j), (m)), temperatures (l), changes in the pH by additions of calcium compounds (lime, etc.) to raise it, or sulphur compounds to reduce it ((g), (n)), by inoculations of legume seeds by *Rhizobium* bacteria to improve nitrogen-fixing in root nodules ((e), (f)) inoculation of horticultural composts by other micro-organisms affecting M activities (c), and by trace element additions (g).

Depletions and deteriorations can result from contrary decisions on all these points. The additions at (d) and (g) can be inadequate in relation to the uptake of nutrients (k) and leaching (j), even when account is taken of the releases from soil particles (AFS). The AFS, M, NF, and S activities can be reduced and some of the AFS and S activities made wasteful, or toxin producing, by faulty conditions ((h), (i), (j), (l), (m), and (n)). There can be pollution of the atmosphere or soil ((h) and (n)). Faulty drainage and high acidity (low pH) are common faults.

Decisions about what inorganic fertilizers (including trace elements) to use are made somewhat less difficult because the natural availabilities of all but a few are commonly adequate. Most decisions concern nitrogen, phosphorus and potassium (N, P, and K); and calcium for pH control.

But decisions are complicated because, even if all nutrients are available, roots may not absorb them in those ratios that are optimal for plant growth and development, plant health, quality of output parts, and the health of animals feeding on them. For example, if the nitrogen available is excessive in relation to the potassium, vegetative growth may be too rank, flower and seed or fruit production delayed, and fungus attacks favoured. If potassium is excessive relative to magnesium, 'luxury uptake' of potassium will reduce the uptake of magnesium and so reduce yields and their nutritive value. There may eventually also be unnecessary loss of potassium in drainage water. If manganese is excessive relative to cobalt in the soil solution, the uptake of the necessary trace of cobalt may be too small so that *Rhizobium* activities (NF) are reduced, and grazing and forage plants do not supply enough cobalt for ruminant animals.

Decisions about inorganic fertilizers are complicated also by the many variations on soil over quite short distances – particularly in their drainage conditions, pHs, nutrient availabilities, and past improvements or depletions and deteriorations.

Further difficulties are in measuring all the chemical flows summarized in Figure 4.1. Gross yields of crops from any one site are in practice difficult and costly to secure. The nutrients returned to the soil in crop residues and dung can usually only be roughly assessed.

Subsoils as well as topsoils should be considered. The chemical methods of assessing the AFS, M, NF, and S interactions are not wholly satisfactory. And these interactions are continuous, and affected by changing weather conditions both before and during growing seasons, so that no feasible monitoring system can simulate them precisely, and they cannot be reliably predicted with precision.

The pace of progress in soil science has been substantial but subjective judgements must still be made of the probable results of most proposed improvement measures.

Examples of the difficulties due to year to year differences in weather, and to local variations in soils are shown in Figure 4.2. The wide variations in farmers'

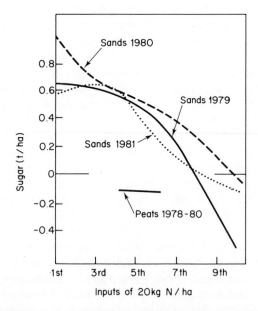

Figure 4.2   Marginal products of nitrogen fertilizer. Sugar beet on two types of soil. Gleadthrope and Rickwood Experiment Stations, England. The products of succeeding increments in nitrogen input varied with the level of total input, soils, and weather in particular years.

actual decisions about inorganic fertilizers, and the apparent low correlation even within one area between outputs and expenditures on fertilizers, are illustrated by Figure 4.3.

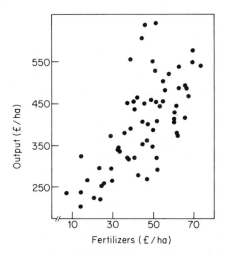

Figure 4.3 Relation of gross outputs to fertilizer inputs (68 farms, Aberdeen area, averages of three harvest years, 1977–9). Farmers differed widely in the outputs they achieved. One measure of inputs was expenditure on fertilizers (including lime). Outputs tended to be higher where expenditures were high. But many other variables also determined outputs.

Fertilizers increase crop outputs more on some sites than on others partly because there are other determinants of crop growth and development. These include all those related to the list (a)–(n) above. The 'green revolution' in the Tropics is based on major increases in fertilizer use, but depends also on major improvements in water supplies and relationships, control of weeds, pests, and diseases, and use of cultivars and timings suited to the improved conditions. And generally, within localities, much depends also on the 'qualities' of individual farmers' detailed subjective decisions about types and amounts of fertilizers.

Thus when considering inorganic fertilizers we can usefully list three groups of determinant of crop output: (i) actual plant nutrient deficiencies that can be made good by fertilizers; (ii) other environmental conditions for crop plants; (iii) 'qualities' of subjective judgements of (i) and (ii). The 'green revolution' and all such developments improve (ii) and (iii) so that nutrient deficiencies (i) can be better corrected. The upward trends of crop yields are due not only to the fertilizers used but also to improvements of (ii) and (iii).

4.9.8 Other flora, and fauna

Poor drainage and high groundwater levels can result, as already noted, in microbial populations that produce toxins and cause loss of nitrogen from soils. Other micro-organisms and higher organisms of many kinds can also increase their populations in soils with damaging effects on crop plants. The most obvious are weeds, insect larvae, and fungal diseases. These are noted further in Section 4.10 below.

4.9.9   Erosion

Water and wind erosion have serious consequences on sites losing topsoil with all its nutrients, structural advantages, and water holding capacity. Water erosion also causes the silting up of reservoirs and river courses and related flood damage, as well as losses to the sea. Wind erosion causes damaging dunes and drifts of sand, and air pollution by finer particles. Although fertile alluvial soils and loess soils have resulted from centuries of erosion in the past, the current gains seem much less than the losses in the areas damaged.

Water erosion can be controlled by the following procedures: (i) keeping soil surfaces covered, and increasing the depth of cover (e.g. using suitable cover crops in the south of the USA; mulching; not overgrazing in savannah Africa; reducing the areas of tilled crops and particularly row crops, and increasing areas of grass); (ii) improving soil structure (Section 4.9.6 above), keeping the aggregates from becoming too small so that water will infiltrate the soil surface and percolate faster; (iii) slowing any movement of water across the surface of land (by stubble-mulches; trash mulches; ridges along contours; use of grass strips along contours; terracing); (iv) careful design of contour ditch and terrace levels, and the run-off channels where run-off is unavoidable. The relation of control measures to local needs and climates (including the probabilities of rainfalls of particular speeds and durations) requires careful planning and design, and raises engineering, farm-economic, and socioeconomic problems (Parts III, IV, V).

Wind erosion can be reduced by measures (i) and (ii) above and by lowering wind velocities near the soil surface by the use of shelter belts, wind breaks, screens, and strips of tall crops such as sorghums and millets. The use of stubble mulches in areas of North America that grow spring wheat is important. Minimum cultivation practices (Section 4.9.6 above) in parts of eastern England can be effective.

4.10   WEEDS, INSECT AND OTHER PESTS, AND DISEASES

The controls of biotic factors damaging to crop plants are of great practical importance. To secure controls of these biotic populations, knowledge of life cycles, and how to break them, is essential. The methods include carefully chosen *crop sequences*, an old form of biological control (Section 4.7 above). *Hygiene* should ensure that infected crop plants or seeds or soils do not spread infection. *Physical* measures at well chosen times control weeds (e.g. stubble cultivations; hoeing early), and heavy shading by some crops is used. Some horticultural soil pests are killed by *soil 'sterilization'* by steam or chemicals. The range and effectiveness of *chemical herbicides, pesticides, and fungicides* are continuously increased, requiring knowledge and skills in use. *Biological controls* are increasingly used, but many depend on adequate understanding of the controlling organisms, and how to make their populations large enough.

Four aspects of control are therefore especially noteworthy, and affect our later discussions (Parts III and IV):

(a)   The importance of *timing*. Unless life cycles are broken at or very near the easiest times, and population increases halted early, difficulties (and costs) are usually greatly increased. Where physical methods are used (e.g. hand hoeing) much hard physical work is required over short periods and some delays are inevitable.

(b)   *Skills* are needed in judging timing in use of chemicals and in biological controls. Research work can define the best times and procedures, but extension education and farmers who will make close observations and act promptly are also necessary.

(c)   Wise controls are therefore *increasingly needed* because many adverse biotic factors are commonly favoured by intensive agriculture – large tillage crop areas, less mixed cropping, more frequent cropping with particular species, genetically more uniform cultivars, more rapid crop growth, heavier output.

(d)   Some control measures depend on decisions about *areas, even regions*, rather than about individual sites (e.g. controls of movements in and between nations of infected seeds or plants; locust control; breeding programmes for resistance to fungal diseases in cereals). Problems in deciding government goals and in organization are therefore raised (Chapters 21 and 22).

4.11   TIMING

Our brief review of crop production and protection has indicated how important it is to be 'in good time'. Most climates provide each year only short periods that are best for particular operations. For optimum yields it is essential *correctly to time* sowing, controls of weeds, pests, or diseases, and harvesting. Soil tillage operations before sowing, early growing season cultivations, and harvest operations are commonly also best timed only within short periods defined by local climates and soil conditions. For example, in areas in the higher latitudes growing spring-sown small grains (such as our Aberdeen, Norfolk, Cortland, and Saskatchewan areas) the starting dates for sowing are limited by low temperatures, wet soil conditions, or snow. The best finishing dates are not considerably later because unless crop establishment is early full use cannot be made of light and temperatures during the growing season, before lack of water and high temperatures slow grain development, or rain or snow cause losses from late harvesting. Figure 4.4 shows how much yields of spring sown barley can be

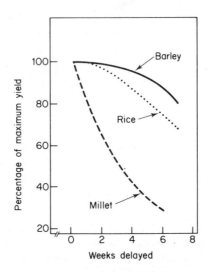

Figure 4.4 Relation of crop yields per hectare to delays in sowing or planting. Barley, Craibstone, Aberdeen, 1960–77; rice, International Rice Research Institute and Pili, Philippines, 1970; millet, Hyderabad, 1973–4.

reduced by delays (Blackett, 1981). In temperate regions where winter wheat is grown the optimum periods for sowing in autumn are limited. If sowing is too early, growth before winter is excessive, using up soil moisture and making plants liable to winter injuries, fungal root rots, and, in some areas, Hessian fly attacks. If sowing is too late, there is winter damage (because the seedlings are not well established), too little tillering, and perhaps late ripening.

In tropical regions with long dry seasons followed by the sudden onset of rains (as in Aurepalle), the correct timings of soil preparations and sowing, and of weed control, are likewise essential to optimum yields (e.g. for pearl millet near Hyderabad) (Randhawa, 1980; Figure 4.4). The effects of delays on rainfed tropical crops are also illustrated by the example for upland rice in Figure 4.4 (Duff, 1978).

Delays also depress irrigated rice yield by amounts determined by the cultivars used, water supplies and relationships, and other variables. New tropical farming systems that depend on careful selections of cultivars for multiple-cropping sequences cannot be successful unless the planned timings are followed (e.g. those for Iloilo).

The optimum timings of harvesting of particular crops are the short periods when they are ripe enough, but not yet being lost by natural dispersal of seed ('shattering'), or too much by bird and other pest damage, or by rain and snow.

In practice, in particular years, the best detailed timing depends on the *weather*, not on long-run averages. Precipitation, temperatures, and even light may be in particular weeks above or below such averages for these weeks. Some climates have especially wide variations. The drier climates of the Great Plains, the steppes, and the savannahs have very variable annual precipitations. The Oceanic climates of north-western Europe seldom have too little water, but risks of excessively wet, cloudy, and cold seasons make timing important in crop establishment, hay and silage making, and grain harvesting. Such unreliabilities of weather require *engineering*, *labour*, and *management* responses (Part III).

REFERENCES

Blackett, G. A. (1981). Optimising spring barley yield. *College Digest*, **1981**, 17–23, School of Agriculture, Aberdeen.

Duff, B. (1978). Mechanization and use of modern rice varieties. In *Economic Consequences of New RiceTechnology*, International Rice Research Institute, Los Banos, Philippines.

Randhawa, N. S. and Venkateswarlu, J. (1980). Indian experiences in the semi-arid Tropics. In *Proceedings of the International Symposium on Development and Transfer of Technology for Rain-fed Agriculture and the SAT Farmer*, p. 213, ICRISAT, Patancheru AP, India.

West, T. S. (1979). *Biosignificance and analysis of trace elements in agricultural soils*. Miller Memorial Lecture, School of Agriculture, Aberdeen.

FURTHER READING

†Bleasdale, J. K. A. (1973). *Plant Physiology in Relation to Horticulture*, Macmillan, London.

Commonwealth Agricultural Bureaux (monthly or quarterly). Abstracts of Publications on *Arid Lands Development*; *Field Crops*; *Forestry*; *Forest Products*; *Plant Nematology*; *Herbage Crops*; *Horticultural*; *Plant Breeding*; *Applied Entomology (Agricultural)*; *Plant Pathology*; *Soils and Fertilizers*; *Weeds*, Commonwealth Agricultural Bureaux, Farnham Royal, Slough.

†Carlson, P. S. (ed.) (1980). *The Biology of Crop Production*, Academic Press, New York and London.

Davis, D. B., Eagle, D. J., and Finnet, J. B. (1972). *Soil Management*, Farming Press, Ipswich.

*Dyke, G. V. (1974). *Comparative Experiments with Field Crops*, Butterworth, London.

Eddowes, M. (1976). *Crop Production in Europe*, Oxford University Press, London.

†Evans, L. T. (ed.) (1975). *Crop Physiology: Some Case Histories*, Cambridge University Press, London.

*Food and Agriculture Organization (1978). *Report on the Agro-Ecological Zones Project*, Vol. 1, *Methods and Results for Africa*, FAO, Rome.

Foth, H. D. (1978). *Fundamentals of Soil Science*, 6th edn., John Wiley, New York and Chichester.

*Greenland, D. J., and Lal, R. (eds) (1977). *Soil Conservation and Management in the Humid Tropics*, John Wiley, Chichester.

Hill, T. A. (1977). *The Biology of Weeds*, Edward Arnold, London.

Holmes, W. (ed.) (1980). *Grass: Its Production and Utilisation*, Blackwell Scientific, Oxford.

Janick, J., Schery, R. W., Woods, F. W., and Ruttan, V. M. (1974). *Plant Science: An Introduction to World Crops*, 2nd edn., Freeman, San Francisco.

Kowal, J. M., and Kassam, A. H. (1978). *Agricultural Ecology of Savanna – A Study of West Africa*, Clarendon Press, Oxford.

Langer, R. H. M. (1972). *How Grasses Grow*, Edward Arnold, London.

70

Langer, R. H. M., and Hill, G. D. (1982). *Agricultural Plants*, Cambridge University Press, London.
Martin, J. H., Leonard, W. H., and Stamp, D. L. (1976). *Principles of Field Crop Production*, 3rd edn., Macmillan, New York.
*Nash, M. J. (1978). *Crop Conservation and Storage*, Pergamon Press, Oxford.
*Simmonds, N. W. (1979). *Principles of Crop Improvement*, Longman, London.
*Soil Survey (1960). *Soil Classification: A Comprehensive System* (7th approximation and subsequent supplements), US Department of Agriculture, Washington, DC.

QUESTIONS AND EXERCISES

1. From a biological standpoint give the principal reasons for the differences in land use for crops in any two of the farm groups in Tables 3.1 and 3.2.
2. In the area you know best, what changes occurred in crop outputs during the last ten years? Explain briefly how biologically they were brought about.
3. How biologically could outputs of the most important crops be increased in the area you know best?
4. In what ways can a skilled, energetic farmer obtain higher crop outputs than are average for his area?

Chapter 5

Animal production and health

5.1 THE PREFERRED SPECIES

As already noted, the main inputs to the animals used in agriculture are feedingstuffs from plants, and unless the plant species and cultivars differ from site to site, the best use cannot be made of climates and soils. From a biological standpoint, therefore, the central problem in animal production is: 'What pattern of animal production systems best suits the pattern of feedingstuffs supplies?'

Throughout human history this has been an important problem and it has arisen even where only rough grazings are available (e.g. in semi-arid or mountain areas). It arises now in more complex forms because feed-grains and other 'concentrated' feedingstuffs can be transported from the areas where they are grown to areas where more intensive livestock production is economic (e.g. our Cortland area; Table 3.4).

The pattern of animal production systems is affected also by climates directly, because these can affect the survival and production of different species and types (e.g. cattle that can thrive in the cool conditions of north-western Europe are unsuited physiologically to the Tropics).

Animal production systems must also be fitted to the patterns of effective demands for their outputs. Whether cattle are used mainly for the production of milk for liquid consumption, as against beef, milk, or butter production or draught power, depends on the geographical pattern of demands as well as on the feedingstuffs supplies (Table 3.3).

The production of milk is based on the physiology and biochemistry of reproduction and early nutrition of young mammals. Egg production should be viewed in a similar way. Meat, wool, and work also depend on reproduction processes for the production of young animals, and on the anatomy, physiology, and biochemistry of these, as affecting their growth and fattening, or wool production, or work capacity.

Matching animal production patterns to patterns of feedingstuffs supply (and possible transfer), climates, demands, and physiological possibilities has involved the use of a wide range of animal species from sponges, snails, and silkworms to cattle, pigs, and elephants. Thus the hill-sheep of north-western Europe, or Colorado, are used to convert poor grazings to lamb (and mutton and

71

wool), but high yielding dairy cows are used to convert less fibrous grazings on lower land, and grain and other concentrated feeds, to milk (with cull calves and cull cows as by-products). And pigs and poultry are used to convert grain and supplementary protein- and mineral-rich concentrated feeds quickly and relatively efficiently into pig and poultry meats, and eggs. Horses, mules, donkeys, oxen, buffaloes, and other species are used for work, each species mainly according to the range of climatic conditions to which it is suited.

A comparatively small number of domesticated species, and variations within them, have made possible the choice of an appropriately wide variety of production systems and patterns (Table 3.3). Some circumstances of climate and feedingstuffs supplies and markets are still leading to attempts to domesticate wild species and to gain fuller control of breeding, feeding, growth, and slaughter of wild species (e.g. red deer, trout, salmon, shellfish). But very large proportions of animal output are based on only a few species that were all domesticated before written history began.

The variations within species are, as in crop plant species, based on differences in genotypes. This is obvious where only one or two genes control easily recognized characteristics such as coat colour in cattle. But many different pairs of genes control the complex biochemistry and physiology that is expressed in reproduction, lactation, growth and fattening, wool and other fibre production, or work (e.g. prolificacies; mature sizes; propensities to fatten at young ages; wool length and fineness; draught power). The selection of breeding animals and controlled breeding have therefore greatly improved the biological efficiency for man's purposes of the species used in agriculture, and the possibilities of fitting animal production systems better to feedingstuffs supplies, climates, and demands. Races and breeds have been selected out and increasingly complex methods used further to improve within them and to secure advantages of heterosis (see Section 5.3.4 below) when they are crossed.

The fundamental importance of nutritional possibilities and problems make discussions of nutrition desirable before genetic improvement is further considered.

5.2   NUTRITION

5.2.1   Knowledge for planning

When animal production systems are chosen, the types of knowledge that are most useful include the following:

(a)   The energy and other nutrients 'required' under defined conditions for the biochemical and physiological processes in maintenance of life, reproduction, lactation, growth and fattening, wool production, or work

(b)   The effects of providing more or less energy or nutrients than are 'required' at any one time (the output–input relationships under the defined conditions and alternative conditions, e.g. in buildings or outside in winter)

(c) The energy and other nutrient values of feedingstuffs available or potentially available (energy, nitrogen in amino acids and other compounds as required for the building up of proteins; major, minor, and trace minerals; and vitamins)

(d) The digestibility of feedingstuffs in alternative diets

(e) The voluntary intake (ingestion) of alternative diets as effected by the appetites of animals and by the palatability of the feedingstuffs

(f) The effects of milling and mixing the components of diets

(g) The effects of grazing controls, the timing of the feeding of concentrated feedingstuffs, and other physical activities and arrangements influencing ingestion and digestion (e.g. free access to roughage feedingstuffs)

(h) The effects of grazing controls (including numbers of animals and timing) on the later productivity of grazing

(i) Which diets result in disease and shorten productive lives

(j) The effects of diets on the quality of outputs (e.g. fat and 'non-fat solids' contents of milk).

Because animals differ genetically and in age, previous nutrition and physical environment, and stage in the reproductive cycle, they differ in their milk and egg yields, growth rates of bone and muscle, rate of fat deposition, wool growth, and work done. They therefore differ in their requirements of energy and other nutrients (a), and, for the same basic reasons, they differ with respect to (b), (e), (f), and (i) above.

The nutritional basis of animal production is thus complex, and animal–feedingstuffs relationships are in practice dynamic – behaviour and outputs in one period being determined by practices and inputs in previous periods.

Nutrition scientists have made most progress in measurements of (a), (c), and (d) above and, in recent years, substantial progress on (e), (f), and (h). Much, however, remains to be measured, and most importantly the effects at (b). The past experiences and personal observations of farmers are therefore built into the detail of production systems that have seemed suitable to their particular circumstances of feedingstuffs supply, climate, and markets. Thus in practice the differences in production systems are wide, and the productivities of systems herds, flocks, and individual animals vary greatly (see Table 11.1). A brief outline of the knowledge built up in (a)–(i) with respect to dairy cows provides examples.

5.2.2 Dairy cows

Nutrient requirements A lactating cow uses energy for maintenance – for basic metabolism (simply to keep alive in the short run), for essential activities in eating and walking, and for stabilizing body temperature. These maintenance requirements are usually regarded as varying with surface area, for which the proxy measure is body weight. She uses energy also in milk production, and this varies with the energy content of the milk. She may use energy also for growing, or for laying down fat (e.g. if she is not yet at mature body weight, or has lost body tissues and fat because of heavy milk yields, or tends for any other reason to fatten) (Figure 5.1). If she is pregnant, she uses energy for the calf fetus and related tissues, and because her metabolic rate is raised. These requirements due to pregnancy rise at an accelerating rate particularly from the sixth month of the nine months of pregnancy. Over the productive life of a dairy cow, these various energy requirements are in total continuously changing, largely because milk production rises to a peak about 6–8 weeks after each calving and then declines until 'drying-off' permits a 6–9 week rest period for the mammary gland before the next calving. As between lactation periods, there are also differences due to maturing and ageing. Delays in conception, abortions, and various diseases can also alter requirements. Moreover, there can be considerable differences between cows in the ratio of energy used to energy ingested because of differences in digestion and metabolism resulting from different genotypes. Thus at any one time in a herd of dairy cows, there are wide differences between cows in their uses of energy.

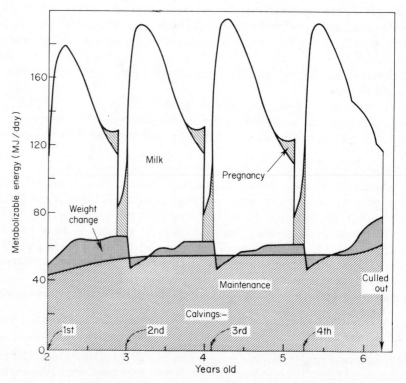

Figure 5.1 Metabolizable energy uses of a cow, as calculated from schedules of 'allowances'.

Uses of protein vary with maintenance needs (and therefore body weight), with milk yield (including protein content of milk), and with fetus development. They are not in constant ratio to uses of energy. Nor are uses of the minerals and vitamins in feedingstuffs. Some minerals are stored in bones or liver so that reserves can be drawn on at times. Some vitamins of the B complex can be synthesized in the alimentary tract but are poorly stored in the body. Vitamin D synthesis depends on sunlight.

Output–input relations   Over the productive life of a cow, the energy, protein, minerals, and vitamins used per 100 kg of milk produced are relatively constant, but the proportions used for milk depend on the ratio of milk production requirements to maintenance and other requirements. This, in Figure 5.1, is the ratio between the white areas (marked 'Milk') and the shaded and stippled areas. The estimates are for a cow of 600 kg live weight yielding 6750 kg of milk of 3.5 per cent butter fat and 8.6 per cent 'non-fat solids' in her third lactation, but culled after her fourth lactation because a fifth calf was not conceived and milk yields fell. Therefore the maximum heights, rates of decline, and lengths of lactation curves are important; also the lengths of the dry periods. Dairy farmers attempt to increase milk yields, and secure regular breeding, with dry periods of not more than about 9 weeks.

They therefore try to judge what would be the results of providing energy and nutrients in greater or smaller amounts. One way is to feed more than 'requirements' in the dry period and after calving until the top of the lactation curve is reached, then according to 'requirements', and later in the lactation according to the 'condition' (essentially live weight in relation to the size of the skeletal frames). The response in milk yields is recorded and changes in live weight assessed usually by subjective 'condition scoring'. But individual cows vary widely: (i) in how they partition energy and other nutrients as between milk production and growth or fattening; (ii) in the shape of the lactation curves; (iii) in their responses to changes in energy and nutrient supplies at different times during dry periods and lactation; (iv) in regularity of calf production (conception rates, and reliable pregnancies and calvings). So judgements of output–input relations on farms are inevitably subjective. They can be improved, but not made wholly objective, by measurements of milk yields and weight gains, and recording of breeding and calving dates, as well as by basic guidelines for energy and nutrient 'requirements'.

Particularly difficult judgements concern: (a) how far live weights should be allowed to fall early in lactations; (b) the effect of energy and nutrient supplies on the number of lactations that can be secured before disease, poor reproductive performance, or low yields lead to culling. Other judgements have to be made about the nutrition of replacement heifers. The use of energy and nutrients tends to be more efficient if growth rates are fast enough, so that heifers can be bred to calve and enter the milking herds at not more than about 2 years old. But experiments have shown that heifers can be fed so much that they are calved only with difficulty and perhaps loss, become less efficient as cows, and have to be culled earlier. On the other hand, heifers can be fed too little in winter for calving at 2 years old.

Nutritional values of feedingstuffs   These depend on the chemical compositions and their effects on ingestion, digestion, and utilization of energy and nutrients in the physiology and biochemistry of animals. Ingestion is best considered after the other determinants.

Table 5.1 sets out some measures for a few important feedingstuffs, including grazing and silage at two quality levels. These data indicate how wide are the chemical differences. The dry matter in seeds (grains) and other plant tissues in which energy reserves have been stored contain less crude fibre and more starch or other easily digested compounds than do the dried vegetative parts of plants

Table 5.1   Measures of nutritional values of some feedingstuffs.

				Energy		
	Crude fibre	Crude protein	Metabol-izable	Net for mainten-ance	Net for growth	Net for lactation
	grams per kilogram of dry matter		*megajoules per kilogram of dry matter[d]*			
UK feedingstuffs						
Grazings						
Poor[a]	288	116	8.4			
Better[b]	155	225	12.1			
Grass silage, clamp						
Poor[c]	380	160	7.6			
Better[c]	305	170	9.3			
Barley						
Straw	394	38	6.5			
Grain	53	108	13.0			
US feedingstuffs						
Grazings						
Poor[a]	390	78	7.8	4.6	1.5	4.8
Better[b]	250	173	11.6	6.7	4.3	6.9
Silage, corn						
Not well eared	320	84	10.2	5.9	3.4	7.1
Well eared	240	80	11.2	6.4	4.1	6.6
Corn, meal	20	100	14.5	9.0	5.9	8.5
Soybean meal	70	496	13.2	7.9	5.2	7.8

Sources: Ministry of Agriculture of England and Wales and Departments of Agriculture for Scotland and Northern Ireland (1975), National Research Council (1978).
[a]In the UK, permanent ryegrass after flowering; in the USA, Napier grass at late bloom stage.
[b]In the UK, rotational grass grazed at 3 week intervals; in the USA, Kentucky blue grass, early vegetative.
[c]Digestible organic matter 52 per cent of dry matter in 'poor'; 61 per cent in 'better'.
[d]4.184 MJ = 1.0×10^6 cal.

(e.g. hay or straw). These vegetative parts become, as they mature, more fibrous and lower in nitrogen content (recorded as crude protein). Vegetative parts also differ because of different species and cultivar compositions (e.g. pastures differ). Water contents differ widely between live, ensiled, and dried materials.

Mineral contents of feedingstuffs vary widely according to soils and uptakes by different plant species and cultivars. The ratios between the mineral elements are complex. Thus some pastures and forages may have important deficiencies if some essential micro-elements (e.g. copper or cobalt) or excess of others (e.g. molybdenum). The macro-elements calcium, phosphorous, and magnesium can be deficient. The proportion of a mineral element that an animal can absorb may be affected by other constituents in its diet, because complexes with other elements can be formed. More of the mineral may therefore be needed in the diet to compensate. Other complexities determine within animals what mineral reserves are stored and what are mobilized, and therefore what is available at any one time for biochemical processes. These complexities can be very important in relation to deficiencies of calcium leading to milk fever (hypocalcaemia) or of magnesium leading to grass tetany (hypomagnesaemia).

The biological values of the vitamin contents are also difficult to assess. One reason is that B complex vitamins are synthesized by microorganisms in the digestive tract and vitamin D by animals receiving sunlight. The availability of vitamin A may be affected by compounds in fungal growth (e.g. in mouldy hay).

Digestibilities These can be affected to a small extent by the pyhsical state of feedingstuffs fed (e.g. whether grains are broken in some way, or grass products chopped), by heat treatments, and by levels of feeding. More important, digestibilities are different in different animal species. In cows and other ruminants, digestion depends greatly on microbial processes in the rumen. These break down plant fibres in ways not important in pigs or poultry, which have 'simple stomachs'. These big differences between species are basic to problems in matching production systems to feedingstuffs supplies. The variations between individual animals within species can be important.

The utilization of energy also differs between species. From rumen fermentations, gases are emitted (principally methane) so that some energy is lost to the atmosphere. In all species, there is also loss of energy in nitrogen compounds in urine. The *metabolizable energy* is approximately 18 per cent less than the *digestible energy* for ruminants, and 4 per cent less for pigs and poultry. But such percentages vary with the compositions of the individual feedingstuffs and the effects of mixing them in different ratios. Furthermore, there are 'costs' in using the metabolizable energy. In calculating the *net energy* available for maintenance of animals or production in various forms, deductions called *heat increments* have to be made from metabolizable energy. These are the losses of energy due to the heat of fermentation (in ruminants) and the heat of nutrient metabolism (in all species). Thus the net energy available from particular feedingstuffs or diets differs according to the species fed. It also differs according to the metabolic processes in which it is used. Thus silage from 'well eared maize (corn)' fed to cattle has a net energy value for maintenance of 6.4. MJ/kg of dry matter; for lactation, 6.6 MJ/kg; for growth, 4.1 MJ/kg.

The value of the crude protein to ruminants depends on digestibility and on the provision of nitrogen for the rumen microflora which build up the amino acids made available later in the digestive tract for the animals. For high yielding dairy cows the supply of these amino acids may be inadequate, so that some direct from the proteins in the diet may be desirable. In such cases the amino acid composition of these proteins may be important: also protection of proteins against breakdown by rumen microflora. If amino acids are supplied in excess, they are used as sources of energy.

Appetites and intakes (ingestion) When planning to meet net energy and nutrient requirements, we should know how much animals will eat of alternative diets. Cows tend to eat more when their energy uses in lactation and pregnancy rise, but the peak of appetite is about 5 weeks later than the peak of the lactation curve, and the correlation of intakes to milk yields is not close. Nor is the correlation of intakes to cow sizes. Diets that are high in digestibility (of organic matter), or metabolizable energy per kilogram (of dry matter), are eaten in larger amounts per day. But, at the higher levels of digestibility and concentration, when physical (gut-fill) controls seem less important in restricting energy intake, physiological controls take over. If more than 60 per cent of dry matter in the diet is in concentrated feedingstuffs as against roughages, and the crude fibre percentage falls below about 15 per cent, the probability of serious reductions in appetite rises. This is because rumen fermentation and other digestive and metabolic processes are altered. Serious metabolic disorders may result and livers may be damaged.

The palatabilities of individual feedingstuffs and diets depend also on physical states. Thus pelleting (cubing) tends to increase intakes particularly of high-fibre diets. The milling (kibbling) or flaking of grain can increase intakes by increasing digestabilities.

Grazing The nutritive value of grazings depends on (i) the total dry matter produced annually by the plants in them; (ii) the climatically based seasonality of productions; (iii) their reliabilities from year to year; (iv) the nutrient and net energy contents of dry matter (which vary with the species composition of swards, rates of growth, and seasonal maturities).

Animals can affect all of these by their grazing, faeces, urine, and hoofmarks, and so in turn be themselves affected. The value of grazing in milk production therefore depends not only on agronomic practices affecting plant growth but also on how and when the vegetation is harvested and the biotic effects of animals. Intensive use cannot be made of grazings for milk production unless (a) they can be divided into areas for heavy grazing over short periods in rotation; (b) surplus grass can be removed at the right times for conservation as hay or silage; (c) seeding of grasses is avoided; (d) physical damage by cows to grass swards is small.

The practical difficulties in meeting these conditions have led increasing numbers of milk producers to bring grass to cows, despite the labour and machinery costs (zero grazing).

Diseases and lengths of productive life   In western Europe and North America, the average dairy cow is culled out after only about four lactations. Much early culling is due directly to inherited propensities to poor reproduction and lactation, but other major causes of relatively short lives are physiological stresses where energy and other nutrient requirements are not well met. Inadequate net energy intakes can result in excessive reductions in body weight and delayed conception. Excessive energy intakes in low fibre diets can cause metabolic diseases. Mineral and vitamin inadequacies can lead to poor performance and disease, including lowered resistance to infectious diseases.

Work, labour, and economic problems   How to match supplies and requirements of nutrients so as to secure the best results in milk production will now be recognized as a biologically complex problem. It also poses many work and economic problems. Different human responses to these problems are indicated by the wide variations in average milk yields from farm to farm, even within particular localities where all specialize in milk production (Table 11.1, Cortland group). Some research work now attempts to measure the effects of less than precise daily provision, cow by cow, of individual 'requirements', so that management and labour problems can be reduced. Tests of mixed 'complete diets' suited to groups of cows that require approximately the same diet specifications are of this type. Other research links problems in nutrition planning with problems in feedingstuffs production, equipment, and labour management – e.g. work on grass silage production (Sections 8.2 and 8.3) and on zero grazing.

5.2.3   Nutrition of cattle for growth and fattening, and of other species

The nutrition of cattle for growth and fattening and the nutrition of other species for milk, eggs, wool, work, and other outputs requires, like that of cows for milk, an understanding of energy and nutrient uses, and a matching of supplies to these. Although basic principles apply throughout, different data and guidelines are pertinent to different species and purposes. Much knowledge has been secured by farmers' observations and by research. Much is still desirable. There is not space enough here to summarize the knowledge and uncertainties even briefly, as for dairy cows. Advisory (extension) publications on practical ration formulation provide useful comparisons between different species and purposes, and actual yield variations provide useful indications of uncertainties.

5.3   INHERENT DIFFERENCES AND GENETIC IMPROVEMENTS

5.3.1   Contrasts with plants

Genes on chromosomes control heredity in animals as in plants. Cattle (*Bos taurus*) have 60 chromosomes; sheep, 54; pigs, 38. Within these species, as well as between species, variations are wide in genotypes and the resulting phenotypes. So significant opportunities exist for genetic improvements – as judged from a human standpoint.

But animals differ from plants in ways that slow the pace of animal improvements. Animals cannot show superiorities in photosynthesis. They can only make better use of the products of photosynthesis (feedingstuffs) through better digestion, and through better partitioning, over their lives, of energy and nutrients as between maintenance and forms of production – growth of muscle and bone, deposition of fat, milk production, reproduction, wool. Genes influence the partitioning in many ways. But so too do differences in environment – particularly differences in energy and nutrient supplies, housing, rearing systems, and other results of management. Therefore genetic effects are especially difficult to measure without confusion by environmental influences on both parents and progeny. And this difficulty is aggravated because the number of animals that can be kept in quite similar environmental conditions in individual herds or flocks is far fewer than the number of plants in a plant breeding station and related test sites. Moreover, cattle, sheep, and pigs take longer than annual crops from conception to the time when their economic yields can be assessed. Therefore the effects of selection and controlled mating cannot be assessed rapidly. And some assessments can be made from only one sex (e.g. milk yield, female reproductive performance, and mothering abilities).

After superior genotypes have been selected, vegetative production and self-fertility cannot be used in animal populations. But, through the use of artificial insemination (AI), individual bulls, rams, or boars can sire each year many more progeny than through natural service. Quite commonly, individual bulls sire 20 000 or more calves a year. This can raise the pace of genetic improvement not only through faster multiplication but also because of storage of semen for several years allows better evaluation of sires and more rigorous selection. Superior females can now be induced by use of hormones to ovulate, and their fertilized ova transferred to other females for gestation, but this process is still costly and limited.

5.3.2 Inherent differences that affect yields

Some differences affect appearances (e.g. coat colours in cattle; absence of horns). Genes determining them are in one or a few allelic pairs. Their frequencies in breeds of animals have been closely determined, and they are used to 'mark' their genotypes.

Some serious malformations and dysfunctions are also caused by one or only a few genes (often recessive) and these can be eliminated from populations.

Some few alleles of particular genes are simply dominant or recessive and control the effects of the other genes (through *epistasis*).

But the many desirable attributes of animals (desirable traits) that determine their efficiencies in converting feedingstuffs to animal products, and quality in these, arise each from the genes in many allelic pairs. These are additive in their effects. Thus, for example, the frequency distributions of dairy cows according to milk yields or butterfat percentages in milk, or of beef cows according to weight of calves at weaning, tend to be 'normal' ('bell-shaped') distributions.

5.3.3 Improvement methods

Criteria of selection   The list of frequency distributions to be considered is part of the definition of the desirable traits and can be, like the plant breeder's conception of an ideal plant, long and complex. Essentially the aims are as follows:

(a)   To minimize the feedingstuffs and other inputs used (i) in rearing replacements of breeding populations, in maintaining these, and for the body 'maintenance' of animals that are producing, as against (ii) directly for production of meat, milk, eggs, or other products

(b)   To secure appropriate qualities of output (e.g. milk that is not too watery, meat that is not too fat)

(c)   In other ways to improve the efficiencies of conversion of feedingstuffs into end products

(d)   To avoid risks of failures, or periods of inefficiency in conversion, due to diseases or malfunctioning

Figure 5.1 provides some indication of how desirable such aims are in dairy cows. The energy required to maintain the cow in lactation and dry after calving, plus the energy required for rearing and growth to first calving, was almost as great as the energy used directly in milk production. Therefore high milk yields per lactation, regular calving at 12 month intervals, sufficiently early maturity to calve first at 24 months old, and many lactations per life are all desirable aims. They can all be elaborated to include such attributes as persistence of milk yield during lactation, reliability in conceiving at first service, absence of difficulties in calving, and resistance to mastitis (bacterial infection of the udder). They can be further extended to include butterfat and 'non-fat solids' contents of milk, colour of milk, propensity to rapid growth of calves for beef production, and quality of the meat from the culled cow herself.

Some conceptions of 'ideals' are of dual or triple purpose animals (e.g. cows to provide milk, meat, and draught power).

Animals can be selected for breeding that have genes causing the progeny to be higher up the scale of one frequency distribution than the population of animals from which the selections are made. But the position of the progeny may be, for better or worse, different on other frequency distributions. Thus raising milk yields tends to lower the fat content of milk. Raising the body weights of cows can help to raise milk yields, and growth rates in younger cattle, but also raises maintenance requirements. Increasing egg sizes in laying fowls tends to raise body sizes and therefore maintenance requirements, and to make egg shells thinner. Emphasizing growth rates and rapid muscle developments in meat animals tends gradually to reduce reproductive efficiency and mothering abilities. Even where only one product is the aim, decisions have to be made about the relative importances (weights) to be attached to individual traits during various stages of improvement programmes.

Equally important is the fact that animals and their outputs are the results of the interaction of genotypes and environments. The genes for high milk yield cannot be fully effective if feedingstuffs supplies are not adequate and well balanced, housing and milking well managed, diseases controlled, and breeding times well followed. Nor can environments with particular feedingstuffs supplies be best used without genotypes in the animal population that are balanced and adequate for the purpose. Thus, in wide contrast, the ample feedingstuffs supplies on our Cortland, NY, dairy farms can be best used through large, high yielding cows, well housed in winter. The fibrous and often scarce supplies of feedingstuffs of our Aurepalle group in southern India are more effectively converted by different genotypes able to survive and produce some milk despite the low nutrient level and hot climate. Improvements in feedingstuffs supplies have often to be made before further progress can be made genetically. Otherwise good genotypes are not well expressed and cannot be well selected. Thereafter nutritional and other environmental improvements and genetic improvements should progress together.

Heritabilities   Table 5.2 lists a few important traits in cattle, pigs, and fowls, and states their heritabilities. These indicate the relative importance of genotypes in the total variances of the traits. For example, statistical comparisons of the milk yields of daughters of particular bulls and daughters of other bulls show that some 20–40 per cent of the total variance of yields can be explained by heredity. Because age of dam, season of calving, nutrition, and other environmental factors on particular farms can all influence yields, the groups to be compared

Table 5.2   Some important traits and estimated heritabilities.

Animals	Reproduction	Production	
Dairy cattle	Interval between calvings, 0.1–0.3	Milk	
		Yield	0.2–0.4
		Fat content	0.5–0.6
Beef cattle	Interval between calvings, 0.1	Growth rate	0.3–0.4
		Carcass quality	0.3–0.4
Sheep	Lambs per year per ewe, 0.1–0.2	Carcass quality	0.3–0.5
Pigs	Weight weaned per year per sow, 0.15	Carcass quality	0.3–0.5
Fowls		Eggs	
		Yield	0.2–0.3
		Size	0.4–0.5
		Growth rate	0.4–0.7

have to be strictly defined and corrections made for variations in lengths of lactation and other variables.

The higher the heritability of a trait, the more accurately can selections of genetically superior breeding stock be made and the more rapidly can genetic improvement proceed. The relatively high heritabilities of butterfat percentages in milk, growth rates in meat production, carcass qualities, and egg weights are in important contrast to the relatively low heritabilities of reproductive performance traits (e.g. as affecting dairy cow calving intervals, and weaned pigs per sow per year). But we should note that the measures in Table 5.2 are based on phenotypic variances in particular herds or small groups. The industry-wide phenotypic variation may be much greater, and heritabilities lower.

Selection differentials If the rigour (intensity) of selection of animals for breeding is so great that only those very near the beneficial end of the frequency distribution scale for a trait leave progeny, then the selection differential is high. This is the difference between the average of the trait in the selected parents and the average in the population from which they are selected.

The additive beneficial effects of genotype improvement would tend to be measured by hereditability × selection differential.

Generation intervals These are the average ages of the animals selected for breeding by the times when their progeny are born. The pace of genetic improvement is determined by these intervals as well as by heritabilities and selection differentials.

Spread of the best genetic material The annual rate of genetic improvement in the whole population of a species in a nation or region also depends on (i) how many progeny are sired by the intensively selective males; (ii) how many develop from ova of the best selected females. In dairy herd improvement, the use of artificial insemination can result in 30 000 calves being sired by one bull in a year. Very intensively selected beef bulls can sire by AI many beef × dairy heifers which may be later naturally mated to beef bulls for beef production (e.g. in our Aberdeen area).

Wide use is not currently made of artificial insemination for sheep and pigs. Few ova transplantations are yet made other than in research centres.

Recording and use of records In order to identify the best genetic material, records have to be kept and test measures made. *Pedigree records* show the ancestry of individuals and so indicate the origins of genotypes. *Performance tests and records* provide phenotypic measures of live weight gains and of the yield and qualities of milk, eggs, wool, etc., and reproductive histories and health records of individuals. *Progeny tests and records* provide similar but genotypic measures and also detailed quality assessments of the meat of offspring.

Modern practices can be briefly summarized as follows. *Dairy bulls* are selected largely from progeny test figures – careful comparisons of their daughters' yields as against the yields of their daughters' contemporaries. But use is also made of pedigree records and cow performance records by selecting sons from matings of bulls with high yielding daughters to high yielding cows with long and sound reproductive histories. *Beef bulls* are selected on performance tests, formal and informal, and to a lesser extent on progeny records. *Boars* are progeny tested, some of the littermates being slaughtered to provide measurements of carcass quality. Sows are performance recorded. *Poultry* for egg production are commonly hybrids of mildly inbred strains (some of different breeds), or of intensively inbred lines. Selection is based on performances of strains, or on progeny testing of strains by crossing them with others to measure which are the best complements to each other – which genotypes best 'nick' together.

5.3.4 Inbreeding, cross-breeding, and heterosis

Inbreeding amongst farm animals tends to 'inbreeding depression' as gene pairs become homozygous, and as such depress particularly the reproductive functions. If used carefully, however, inbreeding can secure some superior types that are more homozygous, with high frequencies of beneficial genes. Outstandingly good animals so contributed the foundation genotypes of important breeds. Inbreeding and heterosis are the genetic foundations of modern egg production. In beef and pigmeat production, crossing of cattle of different breeds is also important. Other benefits can arise from cross-breeding. (For example, where feedingstuffs are fibrous, as in the hills in Europe, small ewes can be kept, and mated to rams of the same breed; but for the last of their pregnancies these ewes are mated on better grazing to rams of larger bodied breeds.)

5.3.5 Human foundations

This brief account of major features of genetic improvement in farm animal populations serves to show how important biologically are the activities of mankind in: choosing sets of criteria for selection; observing; measuring and recording; testing under appropriate conditions; carefully comparing results; selecting intensively and persistently; and developing and making good use of AI and other aids to spread good genotypes, and of modern data processors and computers in recording, comparing, and selecting. The improvement of domesticated animals required human decisions before history began. Modern scientific knowledge and aids have substantially increased opportunities, but also challenges to organization, management and widespread trust, perception and precision, perseverance, and willingness to invest for the future.

5.4 REPRODUCTION

In practice birth rates and early mortality rates vary widely. The interspecies differences are of course fundamental, the respective physiological and biochemical processes being related through evolution to the natural environments in which the wild species found their ecological niches (Table 5.3). But the differences caused by the feedingstuffs supplies and other variables partly or wholly under human control are wide. Even within local areas the differences from farm to farm are important, expressing biologically the effects of differences in management. Thus, in our Aberdeen area, the calves born and reared per 100 cows mated can range from 69 to 100. In our Cortland, NY, dairy area the age at first calving *averaged* 29 months; the intervals between calvings, 12.9 months. And some 29 per cent of cows were removed yearly for 'non-dairy purposes', some because their reproductive performance was poor. The human factor is important too in pig and poultry production. Young pigs and chickens are not as well developed as calves and lambs at birth or hatching. Thus perinatal deaths of piglets commonly average 20 per cent, but can vary from 4 to 35 per cent (Table 5.3).

Many different types of failure or inadequacy can lower reproductive and rearing efficiencies. *Genetic inadequacies* can result from inbreeding or continuous selection of less prolific animals and poorer mothers. Thus single lambs with more attractive phenotypes have often been too continuously

Table 5.3 Measures of reproduction (European breeds in temperate regions).

	Cattle		Sheep	Pigs	Fowls
	Dairy	Beef			
Birth rate (live births per 100 adult females per year)	83–98	70–98	60–200	7–20	1000–20 000
Early deaths per 100 live births	1–10	2–18	3–25	4–35	3–35
Oestrus, natural					
Frequency (days)	About 20		16–17	19	[a]
Seasonality	Not very seasonal		Very seasonal	Not very seasonal	Seasonal
Optimum period for conception (hours)	6–20		20–60	12	[b]
Gestation period (days)	273–293		144–150	112–120	21[c]
Parturitions					
Age at first (months)	23–36		12–24	11–24	5–7
Interval between (months)	12–15		12–14[d]	6–8	[a]
Number in life	3–6	3–10	3–5	2–5	[a]
Surviving progeny per year of dam's life	0.5–0.7	0.4–0.8	0.9–1.8	4.4–12.0	20–400

[a]Periods during which 'clutches' of eggs are laid follow in rapid succession in modern production.
[b]Sperm is stored in the hen.
[c]Incubation period.
[d]Can be reduced by hormone treatment.

preferred to twins, although twinning of sheep is of key importance. Mothering ability has not been fully sustained in some strains of pedigree beef cattle because nurse cows of different types have been used for calf rearing. Although the heritability of reproductive performances is low, continuous selection against high performance tends to become permanently depressing. Many failures result from abnormalities in the functioning of the *endocrine glands*, which control production of ova and sperm and all related sexual reproductive processes, and closely condition behaviour. Inadequate *nutrition* can reduce ova (and sperm) production, cause perinatal and later deaths, and poor growth in young stock. Various micro-organisms cause *diseases* such as brucellosis, resulting in abortions. Other failures are due to abnormal *anatomy* (e.g. small pelvic channels for the passage of calves or lambs at birth). Others are the result of *accidents* or a wide variety of other reasons. *Environmental conditions* (e.g. high temperatures in the Tropics) can reduce the urge to mate. In some intensive production environments, young males require patient guidance before they can serve the females effectively. Even more important is the need to ensure that natural or artificial services are available and given within a few hours in each cycle of ova production in cattle, sheep, and pigs when the placement of sperm has good probability of fertilizing ova. Many production systems rely on *human observation and decision* about the timing of service. The difficulties are considerable, particularly in dairy herds in winter. *Conditions before and after parturition* should be quiet and hygienic, with skilled assistance if required.

The ways of improving reproduction rates are as many and varied as the causes of their depression. Much depends on human observation, knowledge and skill, action and foresight. These are the basis of 'stockmanship', without which biological results are inevitably less favourable.

Modern scientific advances have increased opportunities to secure better results, particularly through better nutrition; disease control; better perinatal care; improved artificial insemination of dairy cows; better care and environments for males, and use of hormones to control ovulation dates; and genetic improvement, including intensive selection pressures in large groups of co-operating farms. An obstacle to improvement is the difficulty of securing sound measures of reproductive efficiency. Substantial additional efforts in recording are required, especially in herds and flocks on range grazings, and in intensive systems of piglet production.

5.5   PHYSICAL AND SOCIAL ENVIRONMENTS AND ANIMAL BEHAVIOUR

For biological success in animal production, the genotypes used must be well suited not only to the feedingstuffs supplies but also to other features of physical environments. The animals must also by instinct and experience develop behaviours that reduce physical and social stresses, and promote individual well-being. Therefore sound development of agriculture requires increasing attention to animals' physical environments and behaviours. Also public demands on moral grounds for protection of animals' welfare are increasing.

5.5.1 Physical

Temperatures The basic physiological relationships are those affecting body temperatures. Within certain ranges of environmental temperatures and humidities, particular species, breeds, and ages can keep their body temperatures almost stable without using much energy, or losing appetite and reducing milk or other production. But as temperatures rise above or fall below these ranges – the 'comfort zones' – there are increasing discomforts, costs, and losses. Table 5.4 gives some examples. These show particularly the vulnerability of young animals

Table 5.4 Body temperatures, comfort ranges, and examples of effects of air temperature changes (European breeds).

	Body temperature (°C)[a]	Comfort range (zone) (°C)
Dairy cows	38–39	−1–22
Growing cattle	38–39	−4–17
Calves	38–39	4–22
Pigs		
Youngest	39–40	19–29
About 114 kg live	39–40	6–18
Poultry		
Chicks, youngest	41–42	34–36 (brooder) 20–22 (house)
Fowls	41–42	10–22

Effects of reduced air temperatures

Dairy cows	From 15 to 5 °C, requirements of metabolizable energy for maintenance may rise 15 per cent. From 20 to 0 °C, digestibility of dry matter may fall 3.6 percentage points. Further research required, owing to variability
Pigs	Feed required per kilogram of live weight gain rises as temperature falls below 22 °C for pigs of 32–65 kg; 16 °C for pigs of 75–120 kg
Poultry	Feed consumption of fowls rises if temperature falls below 7 °C

Effects of raised air temperatures

Dairy cows	Milk yields fall, in most European breeds, above 21 °C (in Brahman zebu cows, above 37 °C)
Fattening cattle	From 19 to 28 °C the energy required per kilogram of live weight gain is increased by 12–15 per cent
Pigs	Feed required per kilogram of live weight gain is increased if air temperature rises beyond 22 °C for pigs of 32–65 kg; 16 °C for pigs of 75–120 kg
Poultry	Above 26 °C egg production, egg sizes, and shell quality are reduced

[a]Variations may be somewhat greater without harm.

(especially piglets and chicks), and the obstacles to efficient production in high and low temperature conditions. For efficient production in the Tropics and Sub-Tropics, therefore, the genotypes used (e.g. zebu cattle) must result in physiological reactions appropriate to high temperatures, activities producing heat must be minimized, and cool shade and even water for external use to help cooling must be provided. In the high, cold latitudes and at the high altitudes, genotypes that result in wool, hair, and other insulators against loss of body heat are important, along with heavy reliance on the heat of fermentation produced in the rumens of cattle and sheep. Because reducing heat loss can be so important in cold conditions, water on skins and where animals lie should be minimized. Hence the importance of long hairy skins and fleeces and provision of dry lying places and housing. Wind increases heat losses. Hence the importance of shelter. Provision of heat for young chicks and piglets is essential.

When changes in environment are made, acclimatization has to take place. If the changes are not great (e.g. when beef cattle are housed in October in north-western Europe) they may be completed within a comparatively short period.

Humidities Where atmospheres are too dry, animals may suffer from dry mucous membranes and chest infections. Where atmospheres are too humid, body heat balances can be more difficult to maintain, and in buildings bacterial and viral infections are common. Insects and other parasites may cause losses in hot, humid conditions.

Ventilation of buildings is important as affecting relative humidities, infections, temperatures, and contents of dust, ammonia, and other gases. If ventilation is poor, animals lose their appetites and become less adaptable to weather changes, and more susceptible to disease. Production falls. But excessive ventilation can carry away useful animal heat and, like draughts, can chill and lower resistances. Well chosen balances must therefore be secured for various types and ages of animals.

5.5.2 Social

Space and controls Space is essential if individuals are to be contented enough to be fully productive. Some species and strains can be conditioned to greater crowding, but beyond certain limits the integrity of individuals is challenged too often when they are crowded and they can suffer severe depression and injury.

Behaviours In attempting to improve physical environments and herd and flock management, man restricts animal movement by fencing, housing, and other restraints. Even access to daylight may be entirely prevented in pig and poultry houses. The control of breeding also restricts natural freedoms. The results can be improvements in animals' welfare (e.g. by reducing fighting). Or animals' social organizations can be disrupted in ways that lower their welfare and production. Farm livestock tend to settle into linear hierarchies or 'pecking orders' – animal A being dominant to animal B; B to C; and so on. Such organization minimizes fighting and waste of energy. Groups must not be too large, or overcrowded, nor 'strangers' introduced too often. Also in choosing feeding systems, the tendency for 'bully' cows and other dominant animals to get more than their proper shares should be countered (e.g. by using partitions or bars at feeding racks).

Housing and management systems should take special account of perinatal behaviours and dangers. It is particularly important that newborn calves and lambs have their dams imprinted in their behaviours by early suckling and proximity, and that dams have their own progeny (or substitute orphans) imprinted. The periods for this filial and maternal imprinting are in nature limited to a few hours. Piglet deaths from early crushing are common and understanding of sow and piglet behaviours is essential in designing housing equipment and procedures to minimize losses.

Behaviour in grazing can also have big biological consequences. Selective grazing of particular plant species, and over-grazing near watering points, can be very important. In hill and mountain grazing, ewes should be experienced in their own sections. Particular groups are therefore reared to live in particular 'territories'. In intensive grazing, understandings are useful of the following: the rate and times of consumption; group behaviours; defaecation habits; and species attitudes to vegetation made patchy by other species.

Individual behaviours are important in sick animals and animals near to parturition. A few individuals will be odd too in their behaviour in relation to new housing or equipment (e.g. cows that will not use open cubicles (free stalls) in loose housing arrangements). A few individuals are difficult to handle. Some become suddenly dangerous.

Stockmanship includes the ability to ensure that as far as possible, the environmental conditions of animals are always suitable, and that animals develop behaviours that result in their well-being and productivity. While animals should realize that stockmen are in the dominant position, they should be led to submit with quiet confidence.

5.6 HEALTH AND DISEASE

Health is a state of complete physical, mental, and social well-being in which animals thrive. Disease is any disturbance to function or structure that detracts from health. Much disease is seen in high death rates, and low birth rates. Much is not obvious. But in all countries, 'sub-clinical' as well as 'clinical' diseases cause big losses. Some diseases (zoonoses) have special importance because they can affect man as well as animals (e.g. brucellosis, tuberculosis, pork tapeworms). And increasingly the prevention of many diseases is sought for moral reasons.

5.6.1 Nutrition diseases

Nutrition diseases result from diets with inadequate energy, protein, minerals, or vitamins. Such diseases occur most in areas where (i) crop production and grazings are poor (e.g. in semi-arid or cold, wet regions); (ii) mineral contents of plants are ill balanced (e.g. in copper, cobalt); (iii) intensive production systems are not fully supported by analytical and educational services; (iv) errors are made in animal management (e.g. diets are changed quickly and rumen fermentations upset). Some nutritional disease in young animals is caused by poor nutrition of their dams (e.g. muscular dystrophy due to lack of vitamin E). Much nutritional disease is sub-clinical, but causes costly reductions of output.

The basic biological knowledge required to prevent almost all nutritional disease has been secured. But detailed local chemical and other assessments are still lacking in many regions.

5.6.2 Metabolic disorders

Metabolic diseases result from physiological failures that are not caused by malnutrition. One example is acetonemia (ketosis) in dairy cows, when ketone bodies accumulate as the by-products of the incomplete use of fats for energy. Knowledge of the basic reasons for most metabolic diseases is still incomplete.

5.6.3 Genetic abnormalities

Inherited defects are not generally important. The genes from which they result are commonly recessive and can be eradicated from well-bred populations.

5.6.4 Environmental stresses

In intensive production systems the restraints, groupings, housing, crowding, movements, and mating control, and in extensive and semi-intensive systems the temperature, wind, rain, and snow experiences – all these impose major stresses. For some animals such stresses are too great because physiologically, and by instinct and training, they cannot react without loss of health. Large production units tend to be more stressful than medium-sized and small, but much depends on size groupings and care in handling.

5.6.5 Infections

Many different types of living organism cause animal disease. They range in size and complexity from viruses through bacteria, fungi, protozoa, flukes, flat and round worms, to insects. Each has its own ecological niche and life history. Therefore predisposing causes of infections, methods of infection, symptoms, losses, cures, and prevention measures are various.

Many organisms are commonly present and liable to invade animal tissues if the natural resistance mechanisms of animals are weak because of genotype, or weakened by nutritional or environmental stresses. Thus mastitis in dairy cows is due to bacterial invasions of udder tissues and resistance can be lowered by genotype, mechanical injury, stresses from milking machines, poor nutrition, or chilling.

Some infections are by organisms that gain entry to and multiply in animal tissues especially easily (e.g. anthrax, cattle plague, foot-and-mouth disease). The building up of sufficient natural resistances is not judged feasible. Much has been done therefore to eradicate such organisms in whole regions, and special controls are imposed to prevent re-infection.

Other infections will continue. The reduction of disease due to them depends mainly on (i) sustaining natural resistances; (ii) controlling the spread and severity of infections by good hygiene (e.g. for milking cows) and use of drugs (e.g. in de-worming cattle); (iii) augmenting resistances by skilled use of inoculations (e.g. of sheep against lamb dysentery, of poultry against fowl pest (Newcastle disease). Some progress has been made in securing and maintaining individual herds and flocks that are 'specific pathogen free'.

5.6.6 Toxins and poisons

A wide variety of chemicals can cause disease and even deaths. They range from simple elements and compounds (e.g. copper, lead) through insecticides to complex compounds produced by fungi (mycotoxins) and poisonous plants. Many of the damaging chemicals produced by infective organisms are of course also toxins.

Where animals have free choice and environments are not polluted by industrial wastes, damage is very uncommon, but in other circumstances, the dangers of toxins and poisons are one more set of reasons for foresight and care. (For example, in our Aberdeen area, ragwort (*Senecio jacobaea*) is a poisonous weed that is avoided at pasture but, mixed in silage, cannot be avoided by cattle when this is their winter forage.)

5.6.7 Injuries

Injuries allow infections and damage health directly. The causes are many – from nails and wire in forage to faulty handling and equipment, slippery floors, and rough roads.

5.6.8 Predators

Predators may cause injuries and death amongst young animals and stress in flocks and herds as a whole.

5.7 QUALITY IN PRODUCTS AND BY-PRODUCTS

The outputs of animal production systems are, like crop outputs, very varied and changeable in quality. Buyers and consumers have many aspects of quality to assess. Precision is commonly not possible. Preferences differ. Cattle for

92

slaughter provide examples. They vary in age; sex; health; whether bred from or 'clean'; total live weight; proportion that dressed carcass weight will be of live weight; size of joints or 'cuts'; relative ratios of muscle, fat, ligaments, bones, and tendons; distribution of fat in relation to muscle; tenderness of meat; juiciness; flavour; and colour of fat and muscle. There may be damage by warblefly larvae, or by rough handling. Stresses in marketing may have made resting before slaughter desirable because otherwise the meat will be purple. Residues of hormone treatments given to secure more rapid growth may be present. After slaughter, inspection may disclose disease (e.g. tuberculosis) making some carcasses or liver and other offal unfit for human food. Further differences in quality may be caused by differences in the slaughter procedures, cooling, 'hanging' and storage, and cutting-up – particularly the hygienic standards achieved, because bacterial action can be rapid and varied. Still further variation is inevitable between the items actually bought by consumers, because these are particular parts (cuts or 'joints') of the animals and therefore vary for anatomical reasons.

Thus, in cattle and beef, some aspects of quality can be judged quickly by eye if their relation to final consumer satisfaction has been learnt from experience. But few aspects can be quickly measured. And some sanitary and chemical residue aspects require costly tests and controls. Milk tests and controls are now recognized as very important to human health and affect production systems (e.g. in controlling tuberculosis and brucellosis). But the non-visual chemical and microbiological aspects of quality of most animal products require special attention, often from the State (Section 14.3.1).

The quality aspects of wool, like those of cotton, are more measurable than those of foodstuff products. Detailed standards have been defined that are related to machine processing needs and final product demands. But even for wool, classing of the 'raw' product in the first stages in marketing depends on quick personal judgements and therefore varies with personal skills as well as for real quality reasons.

REFERENCES

Ministry of Agriculture of England and Wales and Departments of Agriculture for Scotland and Northern Ireland (1975). *Tables of Feed Composition and Energy Allowances for Ruminants*, HMSO, London.
National Research Council (1978). *Nutrient Requirements of Domestic Animals, No. 3: Dairy Cattle*, 5th revision, National Academy of Science, Washington, DC.

FURTHER READING

Bowman, J. C. (1977). *Animals for Man*, Edward Arnold, London.
Buckett, M. (1966). *Introduction to Livestock Husbandry*, 2nd ed, Pergamon Press, Oxford.
Commonwealth Agricultural Bureaux (monthly). Abstract journals for *Animal Breeding*; *Dairy Science*; *Animal Helminthology*; *Livestock Feeds and Feeding*; *Veterinary Bulletin*, Commonwealth Agricultural Bureaux, Farnham Royal, Slough.

Cole, H. H., and Garrett, W. N. (eds) (1980). *Animal Agriculture*, 2nd ed, Freeman, San Francisco.

Dalton, D. C. (1980). *An Introduction to Practical Animal Breeding*, Granada Publishing, London.

*Falconer, D. S. (1984). *Introduction to Quantitative Genetics*, Longman, London.

*Hall, H. T. B. (1977). *Diseases and Parasites of Livestock in the Tropics*, Longman, London.

Hammond, J., Mason, I. L., and Robinson, T. J. (1971). *Hammond's Farm Animals*, 4th ed, Edward Arnold, London.

*McDonald, P., Edwards, R. A., and Greenhalgh, J. F. D. (1981). *Animal Nutrition*, 3rd ed, Longman, London.

†Maynard, L. A., Loosli, J. K., Hintz, H. F., and Warner, R. G. (1979). *Animal Nutrition*, 7th ed, McGraw-Hill, New York.

†Salisbury, G. W., Van Demark, N. L., and Lodge, J. R. (1978). *Physiology of Reproduction and Artificial Insemination of Cattle*, 2nd ed, Freeman, San Francisco.

*Syme, G. J., and Syme, L. A. (1979). *Social Structure in Farm Animals*, Elsevier, Amsterdam.

*Toates, F. (1980). *Animal Behaviour: A Systems Approach*, John Wiley, Chichester.

QUESTIONS AND EXERCISES

1. Contrast the feedingstuff supplies and animal species and numbers of the Aberdeen, Norfolk, and Cortland example farm groups in Tables 3.1–3.5. Explain briefly the main biological reasons for differences.
2. What are the criteria used in selecting male and female breeding animals in the area you know best? What determines the rate of improvement of the most important species on farms?
3. What diseases depress animal production in the area or nation you know best?
4. In what ways can a skilled and energetic stockman obtain animal outputs higher than average in his area?

Chapter 6

Historical perspective and future problems from a biological standpoint

Our overall view of the biological sub-system helps us now to distil out from human history some wisdom that seems essential for the future. In this process, Toynbee's concepts of 'challenge and response' in human history are useful (Toynbee and Caplan, 1972).

6.1 CHALLENGES

Many challenges have resulted from human urges to survive and to reproduce, because these tend to raise human populations to beyond the limits of locally sustained food supplies. In some periods, climatic conditions have been favourable and lessened the challenges (e.g. when the Ice Age ended in Europe). In other periods the challenges have been severe due to unfavourable climatic changes (e.g. when semi-arid conditions became more widespread in the Middle East), or due to shorter periods of drought or flood causing famines (Section 1.3). Many wars have aggravated the challenges (e.g. the many wars between herdsmen of the rough grazings and crop producers with tilled lands). In general the biological success of *Homo sapiens* as measured by the great expansion of populations has raised enormous challenges (Section 1.2).

We should recognize too that appetites are for more than energy and nutrients. Early in history, man wanted more attractive fruit and vegetables, olive oil, milk, and honey as well as more wheat and barley. And we have noted effective demands now for foods well beyond physiological needs (Section 1.4). Demands have been continuous and increasing for wool, cotton, silk, and other raw materials, dyes, drugs, ornamental plants, and animals for various purposes as well as for foodstuffs.

Not least important have been the inquisitive instincts of mankind. These have led to geographical explorations and, particularly since classical times in Greece, to a search for understanding of life on Earth. Accumulated knowledge has itself made challenges to man's abilities to verify and to apply.

6.2 RESPONSES

Archaeology and history provide descriptions of mankind's responses to these various challenges. About BC 16 000, in Mesopotamia and Egypt, Neolithic men and women began to domesticate animals and sow grain. They saved seeds from selected plants. They prepared ground roughly, and probably weeded to some extent. They observed the seasonal movements of Sun, Moon, and stars, and learnt when best to sow in their own areas. Valley floods were used, and more direct irrigation methods learnt. But the full challenges could not be met by using soils and water better in the first settled areas. Emigration to new areas took place (e.g. from about BC 10 000 to central and north-western Europe). So lessons were learnt about climates and soils and pests. Sites for tillage had to be chosen and crops reselected.

In general, Neolithic men and women were far more successful than their predecessors, and provided the biological foundations for the development of towns and civilizations, first in the valleys of the Euphrates, Tigris, and Nile and later in Europe. These were the foundations on which the Greek and Arab spirits of enquiry could be nurtured and, after the Dark Ages, lead slowly to modern science.

6.3 BEFORE 1840 – THE ERA OF HUSBANDRY

But long before this began to be effective, about 1840, the challenges from increased populations had become much greater. They occurred in China and southern Asia and parts of the Americas, as well as in the Middle East and Europe, because agriculture began in prehistoric times in these regions too. Thus there are good reasons to consider what the responses in agriculture were from a biological standpoint over the entire period from early Neolithic times to about 1840, and for mankind as a whole. Of course, deep studies are lacking of many periods, and examples have to be drawn largely from the Middle East and Europe.

6.3.1 Soil fertility and weeds

In such a sweep of history and prehistory (supplemented by modern studies by social anthropologists) it is easy to perceive two main questions that mankind struggled with: 'Where to farm?' and 'How to maintain and increase crop yields?' Some sites and areas were more fertile than others. Yields varied from year to year, and tended to decline. One type of response was to seek the help of magic and religion. This type was substantially conditioned by the times of Buddha, Confucius, Jesus of Nazareth, and Mohammed, but even now fertility cults and polytheistic fears prevail in much of the tropical world.

The direct responses were essentially as follows: (a) to choose sites for tillage carefully; (b) when yields were unsatisfactory, to shift to new sites; (c) to return to old sites (i) when fertility was naturally restored and (ii) when other sites had become less attractive; (d) when satisfactory sites became scarce in an area (i) to clear and improve for tillage local sites that required more labour or yielded less crop or (ii) to emigrate more permanently to more favourable areas; (e) for some people to adopt livestock herding on rough grazings as a permanent way of life.

Thus we can distinguish in the human story four main types of response: (i) permanent settlement in particular areas and efforts to increase and maintain production there, including crop rotations with longer or shorter periods of 'fallow'; (ii) migration to other areas for settlement: (iii) nomadic grazing of domesticated or semi-domesticated animals (e.g. in the Middle East, Central Asia, Masai country in East Africa, and Lappland); and, in some regions, (iv) continuation of hunting and gathering (e.g. by Eskimos; pigmies; the aborigines of Australia).

These types of response were not always and everywhere entirely separate. Some hunting and gathering continued everywhere. Some nomads cultivated a little. Some migration could be regarded as shifting cultivation with long bush fallow periods within a wide ecological area (e.g. in Sukumaland and Geita in Tanzania). But the classification is useful because it quickly indicates how relatively unimportant overall types (iii) and (iv) became, and how important were (i) and (ii).

We should note, however, that the responses in nomadic grazings were more intricate than is often recognized. What is now tradition had to be built up from many experiences of climate and weather, plant growth, animal production and reproduction, and pests and diseases. Natural selection affected the herds and flocks, particularly because nutritional levels were often low; but man also selected so that more useful types of animal were secured quite early. Not least important was the evolution of human social organizations and behaviour to suit the biological conditions. That these responses led to population increases, and an urge and ability to conquer more settled areas, is well known through the Old Testament of the Bible, and stories of Jengis Khan and his nomadic horsemen. But the fuller development of civilization required more intensive production than the biological resources of semi-arid or cold areas could sustain, even when human responses there were great.

The responses in migration for settlement of new areas began as early as tillage and seed sowing, because mankind was already moving about in his hunting and gathering. The necessary size of resettlement in the long sweep of history can be imagined from the fact that grain yields per hectare increased in the 12 000 years to 1840 by perhaps some four or five times but the human population in the world by perhaps 140 times or more. In many ways, the biological processes could be considered as those of a species well enough adapted in form and function to fit many millions of biological niches varying widely in climate and soil. On a scale never before known among vertebrate animals, *Homo sapiens* was able to increase. By trial and error, in many regions, the niches were found. In more human terms, we recognize that the ability derives not only from anatomy and physiology but also from all that '*sapiens*' indicates – from abilities to observe; try; remember and interrelate; foresee and manipulate; and the willingness to forgo present for future satisfactions. The migrants were prepared to bear the hardships and uncertainties of resettlement, but they carried with them skills, materials, tools, crop seeds, breeding animals, and, even more important, abilities to observe and adapt to new conditions. That these were not always enough is evident from many failures (e.g. of Chinese settlements in drier areas; of various prehistoric tribes of Britain; of Norse settlements in Greenland; of many tribal movements in Africa).

We should recognize too that many false conclusions were drawn from particular experiences. Everywhere therefore false beliefs tended to accumulate as well as useful knowledge. In our Aberdeen area, for example, it was believed for many decades that weeds were sown and planted by the Devil, and he could be induced to avoid particular farms only if they set aside a 'Good man's plot' for his own use.

In the end of course the abilities and energies required in the new areas were essentially the same as those in the old. Biologically the most important were those affecting the improvement and maintenance of soil fertility and, closely related, the control of weeds. The decline in fertility and increase in weeds were countered mainly by (i) shifting cultivation; (ii) bare fallows; (iii) hand and hoe weeding, usually with the dead weeds left on the tilled soils; (iv) flooding for a short period; (v) legume and grass 'breaks' in cropping with cereals; (vi) use of 'natural' fertilizers, particularly on gardens and household plots, but also on selected other sites. The human stories about all these are long and fascinating. They have not finished. We can discern, however, that as human populations became heavier on land resources, responses (i) and (ii) became less valid and (iii) and (iv) more and more important, with opportunities in (v) and (vi) recognized in some regions. The Nile floods provided soil nutrients as well as water. The irrigation of rice padis helped to control weeds as well as provide water. The Chinese used night soil, legumes, and irrigation before Western history began. The Romans in the days of their empire used lime and legumes, although they did not know the scientific explanations. Some American Indians put fishes below their maize seeds. The household plots of many African tribes have fertility sustained at a high level for bananas and other selected species. The migrant Fulani have for many decades brought cattle to dung on sites for tillage by Hausa farmers. The stories are particularly fascinating where the challenges were acute. (The permanent cropping systems of the Chinese and Japanese were impressive examples of man's abilities.)

One period of special interest was the hundred years before 1840 in England. The population was rising fast and the Industrial Revolution was beginning. Demands for more food and wool were increasing. Much was done to raise outputs. Weed control was by hoe and harrow, helped by improved crop rotations (and some bare fallows). Much use was made of organic fertilizers, and grass and clover were used in crop rotations. But soil fertility and crop yields could not be raised enough. Adam Smith was led to favour international trade as a solution; T. R. Malthus to favour restraints on population increases. In agriculture there was much searching for helpful new understandings. The Board of Agriculture in 1803 appointed Humphrey Davy as scientist. But not until J. B. Lawes began chemical work on fertilizers at Rothamsted does it become historically convenient to consider that a new era started in the biology of agriculture.

6.3.2 Water management

The search for sound water management in relation to soils runs parallel to that for soil fertility and weed control. Early and widespread advances were made in irrigation in the Nile, Tigris, and Euphrates valleys, in eastern and southern Asia, in North Africa by the Romans, in Spain by the Arabs, in the Americas by the Incas and by the Indians of Arizona. Related advances in drainage had to be made in these areas. In addition, as population pressures increased, there was reclamation of many wet areas. By the thirteenth century, the basic drainage system for Holland was already established. But it was not until 1843 that cylindrical clay pipes began to be used for underground field drainage in Britain, although for many decades before that stones and faggots were used to keep field drainage channels running.

Heavy clay soils were not cropped until much labour was available (e.g. in the Yangste valley) or until ploughs and other implements and draught animals made the work fast enough. Even then, soil structure problems remained especially difficult on these soils (Section 4.9.6).

The breakdown of crumb structures due to continuous cropping was probably one of the main reasons for falling yields and the search for fertility was in part a search for ways to maintain good soil structures.

6.3.3 Climates

After prehistoric men with their irrigation methods, Greeks and Romans seem to have been the first to improve micro-climates. They 'forced' vegetables using glazed frames in winter. Use of glass and hot-beds of fermenting horse dung was successful in Holland and France by the thirteenth century. Glasshouses were built in the late sixteenth century. But in general, crop species and strains and agronomic methods were chosen to suit various climates, rather than climates modified.

6.3.4 Choice of crop species and cultivars

Onions and beans appear to have been cultivated before man could write. The Egyptians had grapes for wines and many herbs for medicine by BC 4000. By Roman classical times, many species were used, including many vegetables and fruits, and beans for green manure. Certainly by 600 AD, the Chinese were drinking tea. Migration and trade increased the range of choices and stimulated the type of widespread small scale trials that farmers and gardeners can make and closely observe, even though they may not measure precisely. Intercontinental movements of such important species as rice, potatoes, maize, tobacco, ground nuts, and many more affected agriculture in most regions. Progress was delayed in many ways (e.g. in Europe in the Dark Ages). But by 1840 everywhere the range of species was already so wide that later adoptions have been few, except for ornamental horticulture.

Mankind was also continuously selecting strains within species. Agronomic conditions may have induced higher mutation rates (Section 4.3) and these increased opportunities for selection of sub-species, strains, and many cultivars (e.g. of *Brassica oleracea*).

6.3.5 Agronomic practices

Man's response also included the accumulation of experiences of how to grow his various crops. Much was recorded and disseminated by classical Greece and Rome. The monasteries wrote down many ideas and, after the Dark Ages, spread them. Chinese ancient literature on growing crops was extensive. By 1523 in England, Fitzherbert of Norbury, Derbyshire, wrote 'The Boke of Husbandry'. But much of great importance for particular localities was never written down.

6.3.6 Animals

In a brief telling of the human story the central importance of animals to nomadic peoples is clear. But settled peoples have in many ways given priority to meeting challenges in crop production, and animal production has followed on, depending on the crops produced and rougher feedingstuffs available on the uncropped land. Thus the design of intensive animal production systems did not generally progress very far before 1840. Milk there was, and wool and meat, but little intensively produced. Draught power and dung were often as important. The monasteries in Britain had many meatless days, but they gave low priority in the production programmes to animals. The beef of manorial times was largely from culled cows and oxen.

In the hundred years before 1840 in Britain, genetic improvement was made in cattle, sheep, and pigs. Feedingstuffs production had been increased sufficiently by improved rotations with turnips and clover and grass, and there were areas of useful permanent grass now enclosed and with grazing better managed. With increased demands for meat and wool the stage was set for clever selective mating to secure breeds. The science of genetics was not available to help, but the lack of it seems to have been felt less than the lack of the sciences dealing with soil fertility.

6.3.7 Pests and diseases

Without modern scientific understanding of pests and diseases, progress in control was limited. Rotations, seed and sire selection, and slaughter of ailing animals provided some controls. Some helpful remedies were also gleaned from experiences in much the same way as in 'folk medicine'. But pests and diseases held production down still further below the low levels determined by poor soil fertility and animal nutrition.

6.4 THE ERA OF 'SCIENCE WITH PRACTICE' – 1840–1975

It was clear before 1840, and the famine in Ireland in 1845–6 confirmed it in more minds, that all the past achievements were not enough. Fortunately the mental attitudes to enquiry had at last developed enough to support a chemical approach.

By 1843, soluble phosphates were manufactured from phosphate rocks and mined nitrates and potash salts were in use. The now commonplace nitrogen, phosphate, and potash (NPK) of 'artificial' fertilizers became available to many farmers in Europe. Before 1914 a German chemist, Haber, knew how to fix nitrogen from the atmosphere. Together with the continued use of lime, the foundations for large and sustained increases in crop and grassland production were laid. So great were the new possibilities opened up that even now they are far from being fully exploited (see Figure 4.3 or Tables 11.1, 12.5, or 12.6).

Chemistry contributed later increasing understanding of animal nutrition and of food- and feedingstuff values. Genetics began to be used in plant improvement, and later in that of animals. Microbiology gave the reasons for the use of legumes in crop rotations, and the causes of many diseases. Veterinary medicine and hygiene became much more effective. Agricultural zoology showed how to control many pests. Later, agricultural chemistry discovered modern types of herbicide. Agronomy and animal husbandry became much more scientific in their methods. Biometric statistics became indispensable. And so on. A detailed list of all that has been accomplished since 1840 would fill several books. Recognition of this may be made easier if the brief lists in Table 6.1 are considered in relation to a few localities and the detailed changes there over the last 10 years. And we should note that the accomplishments inevitably required many difficult marriages of science and practice – of new understandings and old experiences on particular sites. Simple measures of the achievements as reflected in crop and animal production, including changes in a recent decade, are given in Tables 1.9 and 1.10. For the whole period 1840–1980 one reflection is the increase in yield of wheat in England from about 1.9 to 5.1 tonnes per hectare. In Japan, rice yields rose from about 1.6 to 6.3 tonnes per hectare. In the USA, maize yields rose from about 1.5 to 6.4 tonnes per hectare. Another reflection is that the human population in the world increased by 300 per cent. In some regions, human diets improved (e.g. the USA, the UK, western Europe). But in some they deteriorated (e.g. in India and China). When the decades since 1840 come to be fully assessed, the impact of the agricultural sciences on human biology will probably be recognized, along with the settlement of new lands, as explaining much in human history.

Table 6.1 Important variables in the biological sub-system.

1 Natural endowments of individual sites	2 Human inputs[a]	3 Outputs and output–input relations[b] as determined, site by site, by columns 1 and 2 and the following:[c]
Climate	Irrigation Shelter Shading Grass, polythene Heating	*Changes in natural endowment flows* Weather
Water	Drainage	Season
Topography	Flood control Erosion control	Year to year
Soil	pH control Organic fertilizers NPK and other fertilizers Trace elements Tillage and seedbed preparation Mulching	Trends *Human inputs* Timing[d] Design Detailed skills
Natural vegetation	Clearing Grazing control Felling, and other extractions Crop plant species Crop plant genotype improvement Agronomic practices Crop sequences Mixed cropping Spacing	in use Other qualities *Qualities of output aimed for*
Other biotic factors Weeds Disease organisms Pests	Biotic controls Harvesting Other Animal species Animal genotype improvement Animal nutrition controls Animal health controls Animal housing Stockmanship	

[a]Particular types of input may be related to more than one of the types of natural endowment.
[b]Input–input and output–output relations are so determined also.
[c]The relationships are commonly *jointly* determined by the levels of more than one input, as well as by the natural endowment flows.
[d]Including that of past investments, current inputs, and improvements for the future.

6.5   THE PRESENT AND FUTURE PROBLEMS

So it is that as we survey our problems, in the early 1980s, we value highly for the future (a) knowledge and skills accumulated year by year and site by site through the centuries: (b) the attitudes of mind that have resulted in such accumulations and their effective use; (c) the scientific knowledge of the biological sub-system of agriculture that has been built up and verified; (d) the scientific approaches that can add to this knowledge, and help to make its use effective, site by site.

But we can also see that progress since 1840 has added problems, and some have been aggravated by the control of human death rates made possible by the progress of medicine (Section 1.2).

Human population pressures are now greater than ever. In relation to natural endowments of climate and soil they are also uneven. They are rising much faster in some countries than in others, so making this unevenness yet more awkward. Human nutrition is unsatisfactory: millions of children die, and millions do not develop to full health. Millions of adults in the Tropics do not have enough energy for a full life. Millions in the richer countries jeopardize their health by over-eating. The new types of agricultural knowledge derived with the help of science since 1840 have probably been used too unevenly. *Settlement of new lands* has still to proceed (e.g. in south-eastern Asia, South America, and parts of Africa) without full help from the agricultural sciences (e.g. the contributions of soil science, agronomy, and plant protection in defining probable and possible average crop yields and their year to year variabilities).

Soil fertility has declined in some areas (e.g. in thickly populated savannah areas in Africa and drier areas of southern India). Some aspects of soil fertility have deteriorated elsewhere (e.g. organic matter and available nutrients in prairie soils, soil structure in the Indus plain of Pakistan). Soil erosion is serious in some areas (e.g. mainland China, Mexico). NPK recommendations are not yet well defined in very many regions. While the importances of trace minerals for plant and animal production are now known, the complexities due to wide variations in soils are far from being well enough understood and measured to guide agricultural practices closely.

Plant improvements are still being made very rapidly by science-based methods, but in some areas progress is slow (e.g. in improving tropical grasses for animal feedingstuffs). Also there are already signs that the pace of improvements in some areas has fallen seriously (e.g. in wheat and barley breeding in western Europe). There is growing concern that the efficiency of photosynthesis per unit area of leaves must itself be improved, but it is not clear how this can be done.

In *genetic improvement* of fowls there has also been a slowing up in western Europe and North America, and the problems in speeding up scientific improvement of cattle, sheep, and pigs now appear greater than was expected in the 1950s.

For many areas, detailed recommendations on *animal diets* are still not possible because measures are not available of the values of local feedingstuffs, and of the responses of animals over their whole lives to different diets.

Intensive livestock housing to reduce costs of animal products has caused growing concern about the *welfare* on 'factory farms' of poultry, pigs, dairy cows, and calves. Also there is growing realization that the biological efficiency of animals is affected by interplays between their social organizations and human controls and attitudes. Stresses arising from these interplays are not sufficiently understood.

Weeds, diseases, and pests have become more difficult and costly to *control* in some areas despite new armouries of chemical weapons against them (e.g. some permanent weeds in the Tropics; fruit tree pests in North America). The dangers of serious losses may be greater in some circumstances because (i) specialized cropping (even monoculture) is more widespread and intensive; (ii) populations of disease and pest organisms can become resistant to particular chemical controls; (iii) natural enemies of pests are destroyed by chemicals; (iv) mutations in disease organisms can intensify and extend their attacks.

The increased use of chemicals in non-agricultural industry as well as agriculture, and the general rise in human populations, are causing more and more *pollution*. Problems have therefore arisen for agriculture particularly about (i) pollution of water courses by nitrogen applied in fertilizers in amounts exceeding those retained in fields in soils or plants; (ii) pollution of water courses by animal faeces and urine, and effluents from agricultural processes (e.g. silage making; vegetable processing); (iii) use of fields to absorb more of these productively, or at least without adverse effects, and related storage and treatment before application to fields; (iv) pollution of the air and soil by herbicides and pesticides.

In many areas, and especially in the Tropics, the expansion of tillage, the cutting of vegetation for timber and fuel, and over-grazing result in *faster run-off of water* and *drastic alteration of habitats*. Underground water reserves are reduced. Water erosion of soils is speeded up. River channels are deepened in some places and silted in others. Valleys are flooded. Shortages of wood become increasingly serious (e.g. in West Java).

Coal, oil, and gas resources have made possible the use in agriculture since 1840 of new types of knowledge. Coal has provided the energy needed in making and transporting fertilizers; petroleum products have provided that used in mechanical tillage so that work would be timely and difficult soils better managed. Coal and petroleum products have made possible all the transport required as further new areas were settled and trade increased. Such energy requirements are considered further in Chapter 9 but we should note here that fears of *energy shortages* in the future raise substantial biological questions. (Can more soils be managed well with minimum tillage (see Section 4.9.6)? Can crop yields be raised and maintained with less use of artificial fertilizers? Could practices in irrigation, horticulture under glass, pig and poultry housing, and other climatic control practices be altered to reduce their uses of energy?)

Research is now great all along the present boundaries of biological knowledge. But the financial resources for research are scarce. Which sectors should have priority? And for which ecological, demographic, and economic circumstances (Sections 20.3 and 21.3)?

6.6 RESPONSES AND CRITERIA NOW

Thus amongst all the great successes of the era of 'science with practice' various frustrations and fears have in recent years become obvious. They differ in relative importance in different regions, but they all have biological connections. They are moreover increasingly recognized and discussed at international levels, because they have consequences that cannot be kept inside national boundaries. International movements of information, ideas, trade, finance, and people are too great for that.

Most of the frustrations and fears are related also to the work, farm-economic, and socioeconomic systems which are considered in later chapters. These must contribute to our conception of sound development for the future (Chapter 21).

But here the criteria by which future agricultural production systems may be judged *from a biological standpoint* can be tentatively listed, based largely on the fears and frustrations of the present time. Agricultural production systems should (1) be sufficiently productive for mankind; (2) help to reduce year to year variations in production; (3) foster a capability for substantially more production in future; (4) not deplete soil fertility, nor water resources; (5) not increase weeds, pests, and diseases; (6) not deplete unduly forests and other resources of plants of animals; (7) not cause pollution, but help to reduce it; (8) not be unduly dependent directly or indirectly on non-renewable resources (e.g. petroleum products), nor on other production factors that are scarce.

Such criteria will not be satisfied if the use of scientific knowledge and enquiries is not increased and extended. For instance, much more measurement is needed to guide soil fertility practices and animal feedingstuffs production. The many problems of more intensive production need studies in depth. Many more assessments are needed in areas of new settlement, and in any other areas making major changes in land use.

But equally it is clear that more than natural science is needed. All the qualities in *Homo sapiens* will be needed that contributed during the long era of 'husbandry', and since 1840 to 'practice' during the era of 'science with practice'. And in addition other qualities are now necessary. They can best be understood after the work, farm-economic, and socioeconomic systems are considered in Chapters 7–20. What may be regarded as a new era began about 1970–5 – an era of 'systems and strategies'.

REFERENCE

Toynbee, A. J., and Caplan, J. (1972). *A Study of History*, Oxford University Press, and Thomas and Hudson, London.

FURTHER READING

Bronowski, J. (1973). *The Ascent of Man*, British Broadcasting Corporation, London, and Little, Brown and Co., New York, Chapter 2.

Commonwealth Agricultural Bureaux (1980). *Perspectives in World Agriculture*, Commonwealth Agricultural Bureaux, Farnham, Royal, Slough.

Davidson, J., and Lloyd, R. (1978). *Conservation and Agriculture*, Wiley–Interscience, New York.

*Greenland, D. T., and Lal, R. (eds) (1977). *Soil Conservation and Management in the Humid Tropics*, John Wiley, Chichester.

ICRISAT (1980). *Proceedings of the International Symposium on Development and Transfer of Technology for Rainfed Agriculture and the Semi-arid Tropics Farmer*, International Crop Research Institute for the Semi-arid Tropics, Patancheru, AP, India.

*Loehr, R. C. (ed.) (1978). *Food, Fertilizer and Agricultural Residues*, Ann Arbor Science Publishers, Ann Arbor, Mich.

*Loehr, R. C., Haith, M. F., Walter, M. F., and Martin, C. (eds) (1979). *Best Management Practices for Agriculture and Silviculture*, Ann Arbor Science Publishers, Ann Arbor, Mich.

*McEwan, F. L., and Stephenson, G. R. (1979). *The Use and Significance of Pesticides in the Environment*, Wiley–Interscience, New York.

QUESTIONS AND EXERCISES

1. In the area you know best, what practices help to maintain or increase soil fertility, and what practices control weeds.
2. Contrast the soil and other biological conditions on two sites on farms. Choose one site that is tilled at least three years in six, and another that is not tilled so often, or is never tilled.
3. Give examples of seasonal changes in outputs of grass and other feedingstuffs and indicate the related seasonal differences in farm animal numbers, ages, and diets, including use of purchased feedingstuffs, if any.
4. When you judge the farming system that you know best by biological criteria (e.g. criteria (1)–(8) of Section 6.6), in what respects do you find it (a) satisfactory; (b) unsatisfactory?

Part III

The work sub-system

Chapter 7

Introduction

7.1 DEFINITION AND IMPORTANCE

This sub-system is about human work in moving materials, plants, and animals. It includes all labour, and equipment (such as tractors, machinery, hand tools, buildings, fences, irrigation channels) as affecting the movements. But for short it is convenient to call it simply the 'work sub-system'. On farms, human concern with it is deep and often urgent.

Movements on farms are considered in this section. Movements made and energy used elsewhere in processing and manufacture of agricultural outputs and inputs (e.g. nitrogen fertilizers) are not considered here, but their importance is noted, particularly in relation to the use of fossil fuels.

This chapter surveys briefly: (i) the tasks that affect the biological flows; (ii) the physiological limits to human work; (iii) the easing of work. What human work is done in carrying out particular tasks depends, of course, on what equipment is available and how it is used. The relations of labour to equipment and the power to move it are fundamental. Therefore this chapter considers briefly also (iv) the physical and biological aspects of these relations and how they have changed since prehistoric times. Some further examples are given in Chapter 8. Chapter 9 briefly reviews obstacles to change and differences between areas and farms. The farm-economic and socioeconomic aspects are considered in Parts IV and V.

7.2 SUMMARY OF FLOWS

Figure 7.1 helps us to visualize the main interrelations of human work and skills to equipment, and of the work sub-system to the biological. The activities and controls that result in the performance of particular tasks can be provided by different relative amounts of work and skills of farm people, off-farm people, and off-farm materials and power, with different combinations of buildings, fences and other 'fixed' equipment, hand tools, draught animals and implements, tractors and machines. Thus, in hoe farming in Africa, almost all tasks are performed by the work and skills of farm people with simple tools and stores. In great contrast, modern mechanical farming is heavily dependent on tractors and machines and therefore on off-farm work and skills in their making and repairing, and on off-farm materials and power. (See also Tables 9.5 and 10.2 for contrasts between our example groups.)

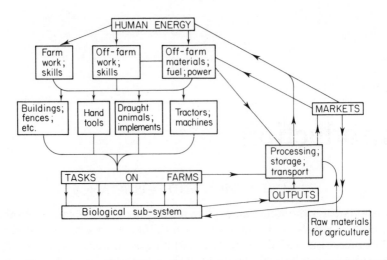

Figure 7.1   The work sub-system of agriculture: A simple diagram of work flows.

Figure 7.1 indicates tasks in processing, storage, and marketing. These increase as specialization and trading increase and more services are demanded by consumers (Section 1.1). In modern economies most of these tasks are performed off farms, with much equipment. Where most farm production is for self-subsistence, the work and skills of farm people, using hand tools, store and process produce for farm household use, and perhaps transport and market some saleable produce locally.

Work and skills, materials and energy are used off farms not only for the manufacture of farm equipment but also for processing of raw materials for other agricultural inputs (e.g. for fixation of atmospheric nitrogen, production of other fertilizers, pesticides, and other agricultural chemicals). The energy needed in this processing is noted in Section 9.2.

7.3 HISTORY

The development of agriculture, and indeed much else in the human story, cannot be well understood unless the work sub-system is well understood.

The design and use of hand tools began when men and women first cleared and sowed patches of ground, or even earlier perhaps when some forage was cut to store for animals. By BC 4000 stones were chipped and then set in horn or bone so as to provide sharp edges in sickle form. Hoes and digging tools led to the use of animals with crude stirring ploughs before BC 3000 in Mesopotamia. The lifting of water using cattle power presented design problems solved in the Arab *shaduf* by about BC 1450.

During the hundred years before 1840 in England, implements for animal draught were substantially improved. Better draught horses were bred. Steam engines first used in pumping water were later used in threshing.

The opening up of the Mid West and Prairies of North America would not have been so fast without the nineteenth century designs of triangular-bladed knife bars, binders for small grain crop cutting and binding, and horse-drawn combine harvesters for cutting and threshing.

The current era of farm mechanization began in the 1890s, when internal combustion engines were first effectively used in field work in the USA. Tractor designs have been greatly improved and the range to suit different circumstances is now wide. Implements have been designed to suit tractors in an ever widening range of tasks. Three-point linkages and hydraulic power lifts became popular in the United Kingdom and North America by 1950. Self-propelled combine harvesters have now greatly increased capacities and reliabilities. Self-propelled harvesters for potatoes, cotton, and other crops have been successfully developed.

Many improvements have been made also in farm buildings, especially for dairy cattle, pigs, and poultry. And, with the use of electric motors, milking, milling and mixing of feeds, wool clipping, and many other tasks are now done more rapidly.

In North America and Europe all these developments have greatly increased the power and equipment used on farms. Few tasks now require much human muscle power. Tasks can now be better timed.

7.4 AFFECTING THE FLOWS OF THE BIOLOGICAL SUB-SYSTEM

When and how well tasks are carried out can greatly affect the biological flows. By increasing rates of work and total capacities for work, mechanization can affect 'when'. And 'how well' can also be affected by the use of more power, by better design, and because timing is better.

Examples for each of the main types of biological flow are indicated in Table 7.1. Thus one improvement of climate-to-plant flows can be by irrigation. For this the tasks in levelling fields and in making field channels commonly have a medium-high power requirement. Timing is not directly important, but high standards of work ensure good flows, minimum seepage along channels, and more even spreading of the water in fields. The tasks are carried out by the use of hand tools (e.g. in Savannah Africa), draught animals and simple implements (e.g. in southern India), or tractors and earth-moving equipment (e.g. in the USA).

Similarly the other examples of tasks in Table 7.1 illustrate how each task poses its particular problems about 'when' and 'how well'; also, how most tasks can be performed by different combinations of human labour, equipment, draught animals, and motors. Three sets of tasks are used to illustrate these basic principles more fully in Chapter 8.

The quality of one type of work done can affect the qualities of later types, and the resulting outputs. For example, seedbed preparation for crops sown in rows (e.g. sugar beet, cotton) should be such that sowing depths and distances can be

Table 7.1 Examples of tasks affecting biological flows and outputs[a].

Flows and example tasks	Affects on outputs of		Power requirement for usual speeds of	
	Timing	Quality of work	Hand work	Machine work[f]
Climate–plants				
Irrigation				
Land levelling and making field channels	L	M	M–H	M
Lifting water	M	L[c]	M–H	M–H
Control of water in fields	H	H	L	—
Soils–plants				
Preparing seedbed for small grain	M–H	M–H	M–H	L
Plant–soils				
Sowing small grain	M–H	M	L	M
Plants–biotic factors				
Weeding				
Physical	H	M	M–H	L
Using herbicides	H	H	—	L
Plants–outputs				
Harvesting small grain	H	M	L–H[e]	H[g]
Plants–animals				
Feeding silage	M–H[b]	L[c]	H	M
Animals–outputs				
Milking cows	H	H	M	L
Manures–soils				
Dung				
Transport	L	L	H	L
Spreading	L	M	L–M	M
Output processing				
Milling small grain	L	L[d]	M	M[d]

[a]L = low; M = medium; H = high.
[b]Depends on feeding system and troughs.
[c]Avoidance of waste is important.
[d]H, if roller mills for flour.
[e]L if heading only; M–H if scything 0.4 ha/day.
[f]The speeds of work, and for field machines their widths and the numbers pulled together, and soil conditions can cause wide variations in power requirements.
[g]Combine harvesters.

precise and germination rapid and uniform. Then thinning and weeding (physically or by herbicidal sprays) can be fully effective.

Most of the examples given provide *opportunities* for better qualities of work, or reduced wastages, when using advanced equipment. But much depends on human observation and deliberate care. Thus elaborate equipment and layouts are *not substitutes* for timely personal controls (e.g. of the flows of irrigation water in fields; of weeding; of hygiene and timing in milking cows).

Close observation and detailed hand work can carry out some tasks much more precisely than can machines (e.g. picking cotton, coffee, tea, or fruit; sorting out fruit or vegetables with visible defects; vegetative propagation in horticulture). Other intricate tasks are more precisely and much more quickly carried out by machines (e.g. sorting of potatoes or apples by size, and, off farms, roller milling of wheat to secure flour of desired 'whiteness'; ginning of cotton).

7.5 OVERCOMING THE PHYSIOLOGICAL LIMITS TO HUMAN WORK

Human muscles are limited in the *work* they can do – that is (forces they can move) × (the distances moved). They are also limited in the *power* they can exert – that is in the rates of work. More power can be exerted for short periods than for long, but even for short periods there are close limits. Important too is the fact that when work is done a large part of the body's energy used is converted to heat. Most of this must be given off; otherwise the body temperature would rise, causing damage. But heat can be given off only slowly if air temperature and humidities are high, and the Sun's rays are hot. The work that can be done is therefore further limited (e.g. in the Tropics). Moreover, malnutrition, diseases caused by infections (e.g. malaria, bilharzia), and lack of physical conditioning to work can all increase the amount of heat to be given off relative to the work done, and so further limit the work that is physically possible.

The importance of these fundamental limitations can be illustrated by the first tillage operations in temperate areas and the Tropics. A fit gardener during cool weather in temperate areas could physically sustain – for 8 hours a day, 6 days a week – a rate of work in digging a good garden soil equivalent to 0.002 ha per hour. A plough with two digger mould boards drawn by a 45 horsepower tractor could do 0.20 ha per hour. The power required to pull the plough is far beyond human capacity. A man and two bullocks with a simple stirring plough can, in southern India, cover 0.02 ha per hour, for some 6 hours a day. In Nigerian dry savannah areas the first hoeing (roughly comparable to the work of a light discer) could be done by an able man at 0.016 ha per hour for 5.5 hours in the cooler parts of each day, during the first rains.

Thus mechanical power derived from fossil fuel (petroleum) and advanced equipment design greatly extends human capacities in the temperate regions. In the Tropics the first hoeings may be shallow and require less work per hectare, but even so the limits to human power are such that only small areas can be covered each day using hand tools. And in practice malnutrition and diseases impose even lower limits.

Some tasks are of course less demanding of work and power than the first tillage tasks. Some harvesting tasks are light (e.g. tapping rubber latex – as against transporting it; pruning of tea bushes and plucking tea leaves). But even so the limits of human capacities for work are fundamental determinants of how agriculture can develop.

We should note particularly that the poorer, less developed regions of the world have the lowest work capacities per head of population because they are the least mechanized, and the most limited by heat, humidity, insolation, malnutrition, and diseases.

We should also note that what can be achieved in agriculture in an area depends on the *total* human work requirement there. In the richer countries, water supplies, sewage disposal, electricity and gas supplies, road and rail transport, security services, communication between farmers and with markets, and other matters do not require much work by farmers. But in poorer countries where the necessary equipment is lacking and service organizations are ill developed, those responsible for agricultural work have themselves to work far longer to achieve services which are far less satisfactory. Thus, for example, many women in southern India or savannah Africa have to spend 1–2 hours a day in walking and carrying to secure water for their households; and the water has many impurities.

In the richer countries the established daily and seasonal patterns of work no longer tend to exceed, even for short periods, the physiological limits to work that can be sustained. If there is stress it is now only for special seasonal reasons (e.g. when harvests are late). But in the poorer countries, with heavy seasonal work, yet little or no animal or motor power on farms, the human power used still tends to exceed the physiological limits that can be sustained for medium and long periods without damaging health. There are in some areas big seasonal changes in human body weights, indicating that health is undermined, particularly when crops are being established and early in growing seasons, following poor harvests.

In all areas, but especially in the poorer countries, plans that would increase the work to be done in agriculture should be carefully examined to see whether the increases make excessive demands on the actually available human power and on human health. If they do, they should be changed to reduce the tasks, or increase the equipment or both. Often adjustments are necessary to reduce seasonal peaks of requirements of human power (Section 12.4.5).

7.6 EASING OF HUMAN WORK

Work tends to increase discomforts and some tasks can be particularly tedious, awkward, heavy, or otherwise unpleasant. The design of hand tools and equipment and the progress of mechanization have reduced the discomforts of human work as well as the human power requirements. Driving a tractor with a well designed driver's seat and cabin results in far more ploughing per day with far less discomfort than walking behind a horse- or bullock-drawn plough, digging with a spade, or wielding a heavy hoe. The use of motor-powered small stationary threshers in southern India not only reduces the human power (and bullock power) needed for threshing and winnowing but also greatly reduces the drudgery and hardship in threshing with bullocks and winnowing by hand and wind. The substantial opportunities to ease human work have been grasped more and more, particularly since the invention of motors.

In recent decades increasing attention has been paid to studies of some tasks as they affect those who perform them.

Ergonomics is the general name given to the subject of such studies. The lifting of weights that is well within human power requires suitable grips, and body posture and movement, otherwise human spines may be injured. Tractor seats and cabins can be poorly designed for the drivers' health. Rapid work in some badly designed milking parlours can be injurious.

As improvements in living standards are seen to be possible, such ergonomic matters become increasingly important, but we should note that skill in the design and use of hand tools so as to ease human work has a long history. This is evident in the tools of the nineteenth century and long before, that are preserved in museums, and in early writings, such as those in China of Chuangtse, who died in about BC 275. His description of how a cook cut up the carcass of a bullock shows early appreciation of fine tools and of skill in their use.

The ease with which tasks can be performed in the work sub-system commonly affects human attitudes to work. This is most evident when mechanization is delayed on particular farms and farmers' sons and other workers leave because they secure satisfactory payments for easier work elsewhere. The use of more or better designed equipment to make tasks easier is commonly regarded as giving 'status' to the operators (Section 12.1.1).

FURTHER READING

*Brandt, J. (1979). Peasant work capacity and agricultural development. Occasional paper no. 55, German Institute for Development Policy, Berlin.

Commonwealth Agricultural Bureaux (monthly). *Agricultural Engineering Abstracts*, Commonwealth Agricultural Bureaux, Farnham Royal, Slough.

*Clark, C., and Haswell, M. (1970). *The Economics of Subsistence Agriculture*, 4th ed, Macmillan, London

Culpin, C. (1981). *Farm Machinery*, 10th ed, Granada Publishing, London.

*Phillips, P. G. (1954). The metabolic cost of common West African agricultural activities. *Journal of Tropical Medicine and Hygiene*, **57**, 12–20.

Preston, T. A. (1978). *Rural Work Science*, Commonwealth Agricultural Bureau Annotated Bibliography No. CAB/96, CAB, Farnham Royal, Slough.

Stone, A. A., and Gulvin, H. E. (1977). *Machines for Power Farming*, 3rd ed, John Wiley, New York.

116

QUESTIONS AND EXERCISES

1. Select two groups of farms that use different 'input mixes' to accomplish their physical work. Describe the 'mixes' briefly and indicate reasons for the differences.
2. Describe three tasks that affect the biological flows and outputs of the farming system you know best. Indicate the effects of timing and quality of performance, and the power requirements for usual speeds of work as in Table 7.1.
3. Give an example of each of the following, and briefly explain: (a) reduction in the power required from farm people themselves for a particular task; (b) use of equipment to make tasks easier in other ways.

Chapter 8

Equipment and skills

This chapter considers three important sets of tasks in order to show how significant are choices of equipment, and skills. Work study methods are then considered. The effects of the layout, sizes, and shapes of fields are noted.

8.1 TILLAGE FOR SMALL GRAIN CROPS AND THEIR ESTABLISHMENT

The essential purposes of this set of tasks are: (a) to provide an environment for the grain crop roots that is physically and chemically favourable, and free of weeds and other biotic factors (Sections 4.9.1 and 4.9.8); (b) to establish the young cereal plants by the most suitable dates, and at the best spacings and depths (Sections 4.1 and 4.5). These purposes may require work on drainage, liming, irrigation, application of dung, control of biotic factors in previous crops, and other tasks. But in this section only tillage and sowing (or sowing and transplanting) and the closely related application of artificial fertilizers are considered. In practice, the conditions for these tasks and therefore their detailed purposes vary greatly. This is evident from the examples in Table 8.1. Crop residues and weeds from previous land uses; soil textures; degrees of soil consolidation; soil water supplies and relations; prospective water supplies; time available – differences in all these affect the detailed purposes. In the following paragraphs, common practices in different areas are described. Individual farmers differ in their practices and would each vary them considerably, according to conditions in individual fields in particular years.

In the *Aberdeen area*, the thick grass turf is ploughed in early winter by tractor ploughs with general purpose mould boards that, when pulled at comparatively low speeds, tilt the furrow slices over about 45° without breaking them much, but ensure easy drainage between them and that all green leaves are well covered. The mould boards are long and convex along the contours, twisting the furrow slices only slowly. The frosts of winter weather the topsoil, particularly that in the exposed angle of the furrow slice. As soon as the topsoil is dry enough in early spring, straight-tined, light harrows till the topsoil and mechanical drills with disc coulters are used to drill in both seed and fertilizer, followed by one rolling. Such operations will usually leave the barley seed in good environmental conditions.

117

Table 8.1 Conditions and special purposes of tasks in establishing small grain crops: examples from five areas.

Example areas	Crop	Previous crop	Soil texture	Special purposes and difficulties
Aberdeen	Spring-sown barley	Rotational grass, 3 years old	Clay loam	Short but thick grass cover to be completely buried; drainage of top soil to be encouraged; weathering by frost
Norfolk	Spring-sown barley	Wheat	Sandy loam	Killing weeds in stubble; avoiding excess drying; consolidation of seedbed
Norfolk	Winter wheat	Potatoes	Sandy loam	Rapid preparation of seedbed
Saskatchewan	Spring-sown wheat	Summer follow	Loam	Killing weeds in fallow but maximizing storage of rain and snow water; early weed control in spring; avoiding drying; consolidation of seedbed
Northern Nigeria	Sorghum	Bush fallow 5 years old	Sandy loam	Bush and grass cover with some thick stumps; dry hard soil at end of dry season; rapid preparation of seedbed when rains start
Iloilo, Philippines	Rice	Rice	Silty clay	Weeds; inadequate water control; if multiple cropping, little time is available

But after some winters disc harrowing may be necessary before the straight-tine harrowing.

In *central Norfolk*, when *spring-sown* barley follows wheat, the wheat stubbles may be cultivated in early autumn so as to control weeds. The equipment used will vary from light, spring-toothed harrows, through heavier disc harrows, to cultivators with broad flat blades – depending on weather conditions and on whether the main purpose is to make the seeds of annual weed species germinate or to grub up and dry (and perhaps burn) perennial weed roots and rhizomes. Commonly a tractor plough with 'digger'-type mould boards will later be used completely to invert the furrow slices, burying all the weeds and debris. The 'digger' mould boards are vertically concave and short with a rapid twist so that they break the furrow slics. The sandy loam topsoil allows good drainage through it and settles well to produce a reliable seedbed with only one harrowing in spring. It is especially desirable to secure good consolidation of the seedbed and avoid excessive drying of the topsoil. So spring tillage is minimized. Fertilizer is broadcast by a separate machine, before seed is drilled in at carefully controlled rates, row widths, and depths.

In *Norfolk*, if *winter* wheat is to be sown after a potato crop, the soil is usually in a much more friable and weed-free state. Ploughing is not necessary, but to break up soil that has been pressed by the wheels of tractors, potato lifters, and trailers, a medium-heavy cultivator may be used, followed by a light roll before the wheat is drilled in. The surface is left a little rough for otherwise it might be weathered down too much and dried to a hard 'cap' in the spring.

In our *Saskatchewan area*, preparations for spring wheat sowing start at the harvesting of the previous grain crop about 21 months before. A main purpose is to accumulate as much water as possible from winter snows, and from rain, and to lose as little as possible by evaporation or transpiration in the fallow period. A common procedure is to cut the previous crop so as to leave the stubble at uneven heights. This reduces wind speeds in the stubble and so traps more snow during the first winter. Summer fallowing cultivations are limited in number but made effective (in controlling weeds and leaving trash on the surface), by careful timing and choice of types of implement such as wide-blade cultivators. Herbicide sprays may possibly be used to supplement cultivations (e.g. to kill some weeds that germinate before the first winter, so that the first spring cultivations can be delayed). The final cultivation and sowing is as early as possible in May, using a discer with seed and fertilizer sowing attachments. Care is taken to sow the seed at a uniform depth of 4–5 cm and to use packer harrows if further seedbed consolidation seems desirable. There may be a light weeding of annual weed seedlings with a rod weeder before the wheat seedlings have emerged too far.

In the *northern Nigerian orchard bush area*, fire is used to remove the thick bush fallow vegetation above ground. However, stumps and thick roots remain, and where these are thick even disc ploughs cannot be successfully used. Moreover the soil is very hard when dry. The common practice is therefore in the dry season only to clear the land by slashing of heavier bushes and fire, and to rush later work as soon as the rains come. Heavy hand hoes are used to loosen topsoil only deep enough for a shallow seedbed. The seeds are dropped by hand behind lighter hoe marks. No fertilizer is used. Practices that use more equipment and less hand labour would require: (i) substantial investment of labour and equipment use in stump and root clearing; (ii) different fallow vegetation (or fertilizers, and more complex cropping plans to maintain soil fertility); (iii) very much more power for working dry soils, or substantial draught power and equipment capable of rapid tillage early in the wet season. Stump clearing by hand, ploughing and other tillage by bullocks with simple implements, and use of phosphate fertilizers and improved rotations are bringing about some changes.

In *Iloilo in the Philippines*, past investments in the levelling and bunding of the small fields have greatly helped in the control of irrigation water and water from rainfalls. Much depends therefore on repair of bunds and on timely blocking and unblocking at particular points. The stubble of the previous crop can therefore be dried out to some extent, depending on rainfall. This helps to control weeds. And when the fields are wet again, stirring by simple ploughs or heavy harrows serves not only to 'puddle' the soil but also to loosen weeds. These are then covered in the mud, or removed if they cannot be controlled by covering. The aim is to provide for the rice seedlings the conditions described in Section 4.9.6. The seedlings may be grown in seedbeds and transplanted, or sprouted seed may be sown (wet seeding). Where the timing of water supplies and other conditions favour it, dry seeding is practised. Where seedbeds are used they are intensively prepared by hand and with close control of fertilizer dressings and water. Transplanting is entirely by hand and requires, like ploughing and weeding, long hours of wading in mud and water, with risks to health from water-borne parasites. The increase in multiple cropping in recent decades is leading to quicker ways of land preparation, weed control and rice establishment (Barlow *et al.*, 1983).

These six examples show how variable are the detailed tasks, the equipment types that can be chosen for them, and the ratios of farm labour to equipment use. They also provide some indications of the importance of equipment design for effective working in particular circumstances. Ploughs, discers, cultivators and harrows, seed and fertilizer drills, and rollers, and at the other extreme axes, cutlasses, and hand hoes, all need careful design. Less evident from this brief survey, but very important in practice, are skills in use of all types of equipment, and the importance of maintenance and repair. These are perhaps most obvious for complex machines with many moving parts (e.g. tractors; combine seed and fertilizer drills). They require regular lubrication, and prompt replacement or repair of worn parts. The tractors require to be well related to the equipment they are powering, and to be driven carefully, so that waste of fuel and excessive wear are avoided. The fertilizer drills need careful attention to limit corrosion. Less obvious but very important are the 'settings' of equipment, because they affect both the power used and the biological sub-systems. For example, the angles and depths of working of discers affect power requirements and the quality of seedbeds. The setting vertically and horizontally of the share points of mould board ploughs in relation to the heel of the land-side affects power requirements, the adequate covering of green material, and eventual uniformity of the seedbed. Even wielding hoes to best effect and with least cost in human energy is not possible without observation and skill (Section 7.6).

8.2   ENSILAGE OF GRASS

In making grass-silage in the Aberdeen area the physical tasks are as follows: (a) to cut rotational grass mixtures from fields of about 5 to 10 ha; (b) to wilt the cut grass in the field to about 25 per cent dry matter for storage in pits or 35–40 per cent for storage in towers; (c) to chop the grass to make it quickly fermentable; (d) to load the grass and compact it in silos, so that air is largely excluded, and fermentation can quickly result in lactic acid lowering the pH to about 4.0–4.5; (e) to minimize loss of energy and nutrients in the grass due to weathering or slow build-up to lactic acid; (f) to secure yields of grass, and digestibilities of silage, that are appropriate to the overall feedingstuffs production and use plans of particular farms; (g) to protect the top and sides of the silage from air, rain, and snow, yet make eventual use easy.

The tasks are related to seasonal changes in grass growth and fibre content. Locally the main period of silage making is usually best restricted to 10 days.

The weather may make the tasks easier (e.g. sunny periods with drying winds may speed the wilting). But the weather may make it impossible to avoid substantial losses (e.g. long rainy periods may delay cutting until digestibilities fall, and may make wilting in the field impracticable). Weather conditions are not reliably predictable for more than 24 hours.

For these weather reasons, important decisions have to be made about the rates at which the various detailed tasks can be carried out. These are: cutting; tossing (tedding) in the field to hasten wilting; picking up and chopping; transport from the field; storing; compacting; and covering. Carrying out tasks at rapid rates requires more equipment and more power, although not necessarily more workers. The greater capability allows cutting to be better timed, more favourable weather to be more fully used, wastage to be reduced, and quality of end product kept higher when weather conditions tend to depress it. Compaction and tight covering reduce fermentation losses by as much as 20 per cent or more of the dry matter in the grass.

Other decisions are about what the quality of end product should be. This quality is affected by the stage of growth of the grass when cut, the rate and extent of wilting, and the thoroughness and precision of the crushing and chopping. Tedding may be omitted, and the forage harvesters used may be 'single', 'double', or 'precision chop'. Where silage of higher dry matter content but also higher digestibility is wanted (e.g. for dairy cows) and where the feeding methods require well chopped material, a capacity for tedding and for double or precision chopping is essential.

Still other decisions are necessary to secure balanced sets of machines and teams of workers. They are based on the total silage to be made and therefore the total grass to be harvested within about 10 days, and the area from which it is to be obtained; also on calculations of the rate of work of the forage harvesters; the time required for transport from harvesters to silo, unloading and return to harvesters; the rate of mowing; and the equipment, procedures, and staff for unloading, spreading, and compacting in the silo. Efficiency depends largely on having a forage harvester of the right type and capacity and on (i) cutting, (ii) wilting, (iii) transporting, and (iv) unloading at rates well related to that capacity, and not resulting in any delays.

Efficiency also depends, as in seedbed preparation, on the design, maintenance, setting, and use of particular pieces of equipment. For example, the design of pit silos themselves and the equipment and procedures should (i) facilitate unloading and consolidation and quick temporary covering nightly to prevent escape of gases and entry of air; (ii) avoid wastage later at the sides and top of silage; (iii) avoid damage to the pit walls by the silage acids; (iv) facilitate feeding of the final product during winter months.

In practice, wide variations occur in the equipment and teams used and in the rates and efficiencies of work (Table 8.2). In a study in Aberdeenshire in 1972, the average rate of flow into the pits was 12.3 tons per hour but the range was wide – 5.1–21.7 tons per hour. The average rate per man-hour was 2.5 tons; but the range was 1.3–4.5 tons. Such wide variations were partly due to the pace of change in machine design. When the survey was made some new designs of mowers and forage harvesters had not yet been fully assessed. Farmers required time to recognize worthwhile changes (Section 20.2.2). And replacement of older machines in good condition might seem premature. But decisions to retain older machines and methods had other reasons also (Sections 11.4, 11.5, 12.1, 12.3, 12.4.1 and 12.4.3).

Table 8.2   Ensilage of grass: equipment used and rates of work (12 farms, Aberdeen area, 1972).

Tasks	Equipment used	Number of farms	Rates of work[a] Mean	Range
			hectares per hour	hectares per hour
Mowing	Knife mower	1	0.53	...
	Flail	3	0.89	...
	Drum	7	1.01	0.40–1.13
	Disc	1	1.17	...
Tedding	Conditioning	1	0.81	...
	Rowing up	1	1.53	...
	None	10	...	...
Harvesting	Single chop	4	0.65[b]	0.20–0.85
	Double chop	4	0.77[b]	0.61–1.01
	Precision chop	4	0.85[b]	0.65–1.21
			tonnes per hour	tonnes per hour
At pit	Rear buckrake			
	+1 man	2	15.9	11.5–20.6
	+2 men	8	10.0	5.0–15.4
	+3 men	1	21.4	...
	Industrial loader	1	17.6	...

Source: Based on data in North of Scotland College of Agriculture (1972).
[a]Actual rates of completion that depended on organization and skills as well as on mechanical capacities (e.g. 'at pit' rates depended on flows of grass to pits).
[b]Where rearward delivery from harvesters required the hitching and unhitching of trailers the rates were reduced by 0.20 ha/hour.

8.3   COW HOUSING, FEEDING, AND MILKING

When cows are tied in stalls and fed and machine-milked there, and the number of cows per worker is no more than 25–35, the rationing of roughages and concentrates can be individual (Section 5.2.2). Close attention can be paid before milking to washing of udders and inspection of the fore-milk for sanitary and udder health reasons. Faults in milking-machine operation and over-milking can be avoided. Lack of appetite and disease can be recognized early, and individual treatments given. If antibiotic treatments are given to individual cows, their milk can be excluded from sales for the appropriate time period. Oestrus can usually be detected and insemination arranged promptly. Inspections for pregnancies can be readily made. Yield and other records can be well kept, and decisions on breeding, drying off, and culling can be well based. Opportunities of 'boss' cows to bully, and the behavioural effects of the comings and goings of individual cows or small groups, can be minimized (Section 5.5.2). The 'atmosphere' in the cowshed can be one of quiet confidence.

When cows are loose housed, and perhaps more than 100 are dealt with by one stockman, the same attentions are still essential to biological efficiency. In practice, their provision depends largely on layout and equipment, and procedures, skills, and care in the milking parlour, because it is here that cows are controlled and closely seen twice daily (thrice on some farms). The rationing of concentrates is usually done here with mechanical aids. Washing of udders and inspection of fore-milk can be well done, and indeed washing becomes important as a stimulant to cows to 'let down' their milk quickly as well as for hygiene. Faults in machine operation can be avoided. Yield records can be well kept. But care must be taken to ensure that high yeilding cows have enough opportunity to eat concentrates. Roughages have to be fed elsewhere and not individually rationed. Sufficient time has to be allowed for all the observations and tasks to be well done, and special efforts made to maintain a quiet confident 'atmosphere'. Methods of identifying individual cows and of communicating information about them from one stockman to another become especially important in large herds.

The effectiveness of work also depends quite largely on tasks outside the milking parlour. Detection of fallen appetites other than for concentrates, and of oestrus, require careful observations. Veterinary inspections and treatments, and artificial inseminations, have to be done – preferably in pens near the exit from the milking parlour. Other important problems are about: feeding roughages; keeping cows reasonably clean; assembly for milking; dispersals after milking; reservation of individuals for special attention; and avoiding behavioural difficulties due to 'boss' cows, and excessively large groups.

In practice, the maintenance of herd health and reproductive performance and the job-satisfactions of workers are especially liable to be inadequate where over 60 cows are kept and the main aim is high labour efficiency. For these and other reasons, the larger the herd and the higher the labour efficiency and returns aimed at, the more important become the design of buildings, the types and design of equipment, and labour skills, observations, and actions. And the more all these should be recognized as closely interrelated.

Careful planning is desirable of work routines because unless all the tasks required for efficiency are completed, and in the right order cow by cow, there are delays, slow milking of some cows, over-milking, and udder damage. Work study methods have therefore been applied to milking operations and have helped to secure high outputs of milk relative to the human labour used without sacrificing biological efficiency. Studies have also proven the importance of training, good communication, and all the attributes of good stockmanship (Sections 20.1 and 20.2).

8.4   WORK STUDY METHODS

Most agricultural tasks are seasonal, and subject to weather and many natural variations. Time and motion studies of them would be much less useful than such studies of repetitive indoor tasks in manufacturing industries.

Work study methods that have been found more generally applicable to agriculture have been called '*work simplification*'. This came to be recognized as important in the USA in 1942 when wartime conditions called for increased outputs with less labour and limited equipment. Work simplification was defined as: 'Better use of available machinery and other physical facilities, to eliminate or combine operations to conserve labour, and to reduce the time and fatigue when hand labour must be used'. Thus in ploughing and harvesting grass for silage, work can be 'simplified' by correct routing of tractors and other equipment. On dairy farms work can be simplified by correct positioning of driveways, gates, and roughage feeding troughs, as well as correct work procedures in milking parlours.

Case studies are required. Work study by experimental methods would often be costly because expensive equipment and layout changes would be necessary. But in general the importance of farm work in human life round the world merits more widespread and scientific investigation.

8.5   FIELD SIZES, SHAPES, AND LAYOUTS

Field sizes and shapes were determined in many regions by topography and water flows. Also, when small farms were cleared from forests, and tillage and harvesting were by human labour and hand tools, or draught animals and simple machinery, field sizes could be small and shapes irregular. Other influences on sizes were length of crop rotations, the value of shelter for livestock in some areas, and in some the need to control wind erosion. The wish to share out among farmers and farmers' heirs land of different qualities (or dates of ploughing in particular years) also influenced field sizes and shapes: hence, for example, the strips of the 'open fields' in England before the enclosures which consolidated and rearranged holdings; hence also the dispersed plots of many Indian and African farmers today.

With the wider and faster machinery that tractor draught has made possible, small and irregularly shaped fields reduce the areas tilled or harvested per man-hour and raise fuel costs. For example, if the machine is 3 metres wide and the tractor is driven at 6 km per hour, a rectangular field of 4 ha would take per hectare almost 40 per cent longer than a field of 20 ha. And if the machine is 5 metres wide, 51 per cent longer. In a 20 ha field, harrows 5 metres wide will till 13 ha per day if the tractor speed is 6 km per hour; but in a 40 ha field, harrows 8 metres wide at 8 km per hour will till 28 ha per day. Irregular fields or fields with obstructions in them may require 15 per cent or more time than rectangular fields with a ratio of 4 : 1 between the lengths of long and short sides (Sturrock and Cathie, 1980).

Small sizes and irregular shapes of fields can also, by requiring more turning of machinery, use more seed, fertilizers, and spray materials – as well as more labour.

On the other hand, where soils differ so much over shorter distances as to make different management advisable, fields should not include different management types (Section 4.9). And where land is used for permanent grazing, small fields may be desirable to reduce the labour in controlling the times of grazings of individual fields, and hedges may be desirable for shelter.

Field boundaries should not, however, require too much labour in maintenance, nor too much land. Nor should they foster weeds or bird damage. The general layout of fields, positions and widths of gateways, access lanes, or driveways, siting of houses and farm buildings – all these must still be related to topography and water flows, availability of drinking water and other 'services', shelter, public roads, size of farms, land-use plans, rotations, and livestock plans. An important influence since the mid 1930s in many countries has been the increasing desire to control water and wind erosion of soils. The relation of layouts to topographies, water flows, and provision of shelter has therefore to be in some circumstances closer than before.

But in many countries, work/labour/equipment possibilities should also be regarded as increasingly important determinants of layouts and sitings. The costs of labour, equipment use, and fuel may differ quite substantially between farms well and farms badly laid out (e.g. in the time required to assemble cows from grazings for milking, or in the transport of roughage feedingstuffs, or bulky cash crops). Where small farms have been amalgamated, changes are needed to provide fully satisfactory layouts for the resulting larger farms.

Layouts, field boundaries, and building designs and sitings affect the appearance of the countryside, and wildlife. They therefore raise human social issues (Section 21.6). Here we should note that, even in comparatively flat areas with little land in permanent pastures, there are commonly opportunities to plant trees, retain some hedges, and protect or create small lakes and ponds. Satisfying landscape can thus be provided and some wildlife preserved, without adding appreciably to annual costs in labour and equipment use, or causing considerable losses of crop or livestock outputs.

REFERENCES

Barlow, C., Jayasuriya, S., and Price, E. C. (1983). *Evaluating Technology for New Farming Systems: Case Studies from Philippine Rice Farming*, International Rice Research Institute, Los Banos, Philippines.

North of Scotland College of Agriculture (1972). *Mechanization systems – wilted silage.* Bulletin no. 4, North of Scotland College of Agriculture, Aberdeen.
Sturrock, F. G., and Cathie, J. (1980). *Farm modernisation and the countryside.* Occasional paper no. 12, Department of Land Economy, University of Cambridge.

FURTHER READING

Barnard, C. S., Halley, R. J., and Scott, A. H. (1970). *Milk Production,* Iliffe, London.
Butterworth, B. (1979). *Materials Handling in Farm Production,* Granada Publishing, St Albans.
Culpin, C. (1981). *Farm Machinery,* 10th edn, Granada Publishing, London.
Raymond, F., Shepperson, G., and Waltham, R. (1975). *Forage Conservation and Feeding,* 2nd edn, Farming Press, Ipswich.
Stone, A. A., and Gulvin, H. E. (1977). *Machines for Power Farming,* 3rd edn, John Wiley, New York.

QUESTIONS AND EXERCISES

1. Select an important set of farm tasks, but not one described in Chapter 8. Indicate the important biological aspects, and describe the equipment, labour, and skills used.
2. Give an example of improvements made during the last 10 years in the design of equipment for some particular farm task. Indicate what has determined the pace of adoption on farms.
3. What have determined the sizes and layouts of fields in the rural locality you know best? What conflicts of opinion about landscape and wildlife as affected by agriculture in this locality are discussed? Can such conflicts be reduced?

Chapter 9

Obstacles, differences, and variations

This chapter considers the obstacles to mechanization, including the potential world shortages of energy to power equipment. Examples are then given of the wide differences between our example farm groups and within those groups in how efficiently work is done. The determinants of such differences will be considered in Parts IV and V.

9.1 PHYSICAL AND BIOLOGICAL OBSTACLES TO MECHANIZATION

In some circumstances and for some tasks, there are substantial obstacles. Field sizes and shapes (Section 8.5) have many historical and social determinants, including all those that determine the sizes, ownerships and tenures of farms (Sections 15.2.1, 15.2.4, 15.2.5, 15.2.8, and 15.2.10). A re-laying out of field boundaries is therefore commonly a slow process in countries with heavy population pressures on rural land and small farm sizes. Moreover the ratio of human labour to the capital and annual costs of equipment is determined also by (i) the number of people seeking to work on farms for their living; (ii) the availability of capital to invest in equipment; (iii) the skills and services available to maintain equipment; (iv) the fuel available to power it. Thus economic as well as social obstacles may be substantial.

We should also recognize that the tasks on some tropical soils being cleared of forest or bush fallows are difficult to mechanize fully (see Section 8.1: northern Nigeria). The removing in one operation of the obstructions of big trunks and stumps without causing excessive soil movement and soil erosion is commonly impossible. The transplanting of rice from seedbeds to wet padi fields has so far not been fully mechanized even in Japan, where mechanical aids have been in common use for two decades. Weed control by physical means between growing crop plants established in rows can be mechanized, but not fully. Some hand work with hoes may be necessary (e.g. in sorghum or cotton crops and where there is much surface trash). Or methods may have to be altered (e.g. to make skilled use of herbicides). Steep slopes cannot anywhere be satisfactorily tilled or

127

harvested annually by modern machinery. Soils that are very stony or in places very shallow above rock outcrops obstruct some types of tillage and harvesting. Much work with permanent crops in both temperate and tropical regions is still by hand (e.g. transplanting, pruning, and harvesting fruit, tea, coffee): mechanization does not secure good biological results and seriously lowers the quality of output (e.g. in the pruning and harvesting of tea). Some useful mechanical aids may, however, be used (e.g. hydraulic lifts and pruning shears to help hand work in apple pruning). And some new cultural systems could be adapted to secure acceptable results (e.g. to allow mechanical harvest of raspberries). In livestock management, even in dairying, there are limits to the support that equipment use can give to hand labour and the human observations and skills that accompany it. (Section 8.3).

9.2 FUTURE SUPPLIES OF ENERGY TO POWER EQUIPMENT

Apart from solar energy used in photosynthesis by plants, about 90 per cent of the energy used each year by mankind is obtained from fossil fuels – coal, oil, and natural gas. These natural resources will not be renewed. How many years they will last depends therefore on (i) how big they are; (ii) whether known and accessible; (iii) the annual rates of use. Important estimates are summarized in Table 9.1.

Table 9.1   Gross energy in fossil fuels: estimates for the world.

	Consumption, 1976	Known reserves	Forecast total reserves
	per cent	years at 1976 rate of consumption	
Oil	48	11	34
Natural gas	19	45	212
Coal	33	1400	2500
Total	100	476	882

Source: Blaxter (1978).

In using these estimates, care should be taken. The total of reserves implies complete and costless substitution between types of fuel: but substitutions will entail losses and costs. Physical access is difficult, especially to many of the known coal reserves, and probably to many reserves not yet found. Legal access is also difficult because the distribution of reserves between nations is very unequal, and differs widely from the geographical pattern of rates of use. The underlying tendency is for the annual use of energy to increase by some 5 per cent yearly so that by 1986 it could be over 60 per cent greater than in 1976, so reducing the apparent reserves.

Thus shortages of fossil fuels, and particularly oil, seem to set important limits to what increases in energy uses should be. If much more energy can be obtained from renewable resources rather than fossil fuels, high rates of energy usage can continue, and even increase. Much research and development work is therefore now devoted to securing energy from nuclear, wind, and water sources, and from the Sun's radiations by physical means. Also as Figure 3.1 shows, the biological sub-system of agriculture could itself be tapped for energy by using solar energy to grow selected plants for fuel. It can also be tapped by burning or fermentation of livestock faeces, straw, and other wastes. In poorer countries, dry dung and straw have long been used as fuels, and use is increasing of simple equipment for 'bio-gas' production by fermentation. We should, however, recognize from Figure 3.1 that if organic matter is used for fuel production and not returned to soils, the fertility of soils will be lower than it otherwise would be, partly because of poorer structure (Section 4.9.6) and partly through loss of nitrogen, and possibly also of minerals if the ashes are not well collected up and distributed (Section 3.2.1).

The possibilities of sustaining or increasing mankind's use of energy will therefore be determined by discoveries of reserves of fossil fuels, substitutions between fossil fuels, research and development for use of energy from renewable sources (including sustainable photosynthesis in agriculture), and by international trading arrangements.

Because there are uncertainties about all of these and because natural reserves of oil are so small relative to annual uses of oil, substantial reductions in the annual rate of total energy use are widely judged to be desirable. And big changes in primary sources of energy used seem inevitable.

This general world position may require substantial changes in energy use and production *in agriculture*, and may be an obstacle to further mechanization. In developed economies, the energy used (apart from solar energy for photosynthesis) in all the processes leading to food on the table is much greater than the energy used as fuel in tractors and other motors on farms. These and other items of equipment have to be made, distributed, and repaired as well as powered. And for the biological sub-system of agriculture, much energy is also used to fix atmospheric nitrogen for fertilizers and in the manufacture of pesticides and other agricultural chemicals (Figure 7.1 and Table 9.2). Energy uses off farms are even greater – for storage, processing, transport, packages and packaging, distribution, refrigeration, and cooking. In total, the energy used far exceeds the energy in the food finally consumed. More and more has been used on farms because of mechanization and increased use of chemicals. And more and more has been used off farms, particularly in food packaging, refrigeration, and distribution. The total energy used for the food system has been increasing fast in the USA, and is some 15 per cent of the total energy used.

In poorer countries, the comparable figures per head of population are low. Work on farms is mainly by humans and animals, and less use is made of chemicals in agriculture. But irrigation and other pumps, and road transport, are substantial users of fossil fuels. For all energy uses some contrasts are set out in Table 9.3 of the energy uses in crop production.

Table 9.2   Energy used in food systems: the USA and the UK.

Main purposes	USA		UK
	1940	1970	1976
	thousand kilocalories per day per person		
All purposes total	14.2	29.3	26.8
	per cent of total		
On farms			
Direct fuel use	12.9	15.2	6.0[a]
Machinery, repairs	3.5	4.7	2.2
Fertilizers, other chemicals	1.8	4.3	5.6
Total	18.2	24.2	13.8
On imports			
Animal feedingstuffs	[b]	[b]	2.6
Human foodstuffs	[b]	[b]	9.1
Food industry	39.2	26.6	23.0
Food distribution	20.9	25.7	19.7
In homes, cooking, freezing	21.7	23.5	31.8
Total	100.0	100.0	100.0

Sources: Steinhart and Steinhart (1974); Blaxter (1978).
[a]Including transport to and from farms.
[b]Included below.

Almost all development on and off farms tends to use more energy that is from fossil fuels, mainly oils. If future development of the poorer countries is to include more irrigation, mechanization, and fertilizer and other chemical use, then the call on world energy supplies will be greatly increased because of the large proportion which the poorer countries make up of the world as a whole (Tables 1.1 and 1.2).

Thus, for the future, many choices have to be made about the development of energy supplies, distribution within countries between agriculture and other users, and between possible uses within agriculture. These choices are essentially economic choices because energy is only one amongst the many scarce factors of production, and saving energy is only one amongst the many aims of mankind (Section 21.8). But it is clear that in considering future plans for the work sub-system of agriculture and for the biological sub-system as affected by it and by agricultural chemicals, special care is now needed over uses of fuels. Examples of practices that could be developed further to reduce energy use include: (i) minimum tillage methods of soil management; (ii) trickle irrigation; (iii) biological controls of pests; (iv) biological methods of fixing atmospheric nitrogen (Sections 4.9.4, 4.9.6–4.9.7, 4.10); and (v) production of selected crops

Table 9.3   Energy used and secured in crop production[a].

Crop	Human labour used	Animal feeding stuffs used	Equipment making and use	Fertil- izers; other chemicals	Total	Energy secured in crop outputs
	hours per hectare	*million calories per hectare*				
Sorghum						
Sudan (1965)	240	0	16	0	16	2 970
Rice						
Philippines (1965)	576	952	41	129	170	6 004
Wheat						
UP, India (1966)	615	2247	41		41	2 709
USA (1971)	7	0	2302	945	3247	7 537
Maize						
Mexico (1951)	1144	0	16	0	16	6 843
Mexico (1951)	383	693	41	0	41	3 312
USA (1970)	22	0	4001[b]	2323	6324	17 882

Source: Summarized from Pimental (1979).
[a]Excluding seed, and off-farm transport.
[b]Including 187 million calories for irrigation.

as energy sources may be extended in some regions of low population density (e.g. Brazil, New Zealand).

9.3   DIFFERENCES IN THE WORK SYSTEMS OF EXAMPLE FARM GROUPS, AND VARIATIONS WITHIN GROUPS

The contrasts in how work is done on farms can be wide because there are many variables in the work sub-system (Table 9.4). Some countries have proceeded further than others in changing from hand labour to use of draught animals and then tractors. The reasons include the physical and biological obstacles discussed in Section 9.1 above. Tables 9.3, 9.5, and 10.2 give examples of wide differences between our example groups.

Between farms within areas there are also important variations, as we would expect from Sections 8.1–8.3 and 8.5. Table 11.1 gives further examples of these variations. They occur not only in complex mechanized farming areas but also where work is largely by hand, or by hand and draught animal (e.g. Iloilo).

From a technical standpoint the efficiency of labour and equipment depends on (i) the design of particular items of equipment (including tools); (ii) the teaming-up of items and individuals for particular tasks; (iii) skills (Sections 7.6 and 8.1–8.3). The outputs of crop and animal products secured depend also on

Table 9.4   Important variables in the work sub-system.

1 Tasks[a]	2 Human inputs	3 Outputs and output–input relations,[b] as determined by columns 1 and 2 and the following:[c]
Flood control Drainage	Tools Machines	Changes in natural endowment flows
Clearing Irrigation	Other equipment	
		Weather Year to year
Glass and shelter control	Tractors	Season trends
Use of organic fertilizers	Other motors	
Tillage	Fuel	Natural and human obstacles
Use of inorganic fertilizers	Animal work	Topography Natural vegetation
	Human work, including	Wet lands
Seedbed preparation	Timing	Hard soils, rocks
Mulching	Maintenance	Field sizes, shapes
Weed control	Repair	Fuel supplies
Spraying	Team compositions Detailed skills	Physiological limits to human work
Harvesting	Work simplification	
Pruning		Qualities of output intended
Animal Grazing control Housing Feeding Breeding Milking Dung handling		
Crop and animal product Processing Storage Marketing Transport		

[a]The peculiarities of each task are determined by natural endowments, human improvements or depletions in the past, and the details of the chosen biological production systems.
[b]Input–input and output–output relations are so determined also (Section 12.4.1).
[c]The relationships are commonly jointly determined by the levels of two or more inputs, as well as by the natural endowment flows.

when the tasks are completed and on all the many variables within the biological sub-system (e.g. choice of cultivars, tillage operations, fertilizers, seed quality, controls of the biotic factors). We have noted the wide variations in biological efficiency within areas (Chapters 4 and 5).

From farm-economic and socioeconomic standpoints, the variations in efficiency have many determinants. These we shall consider in Part IV, so as to be able to judge better what changes and developments should be made (Chapter 21).

133

Table 9.5   Tasks, labour, and machinery: average farms in nine example farm groups, 1978–9.

	Tasks[a]	Total labour used	Machinery use costs	Tasks per man-equivalent	Use costs per man-equivalent
	man-days	man-equivalents	US$'00	man-days	US$
Aberdeen	622	1.9	165	328	8660
Central Norfolk	1159	3.3	353	356	1 0840
Cortland, NY	1539	2.9	390	527	1 3356
Central Illinois	1744	1.2	277	1503	2 3888
Saskatchewan	1704	1.1	130	1508	1 2302
Louisiana	2202	2.8	339	798	1 2278
Iloilo	53	2.3	0.5[b]	28	26[a]
Aurepalle	25	2.3	0.5[b]	25	21[a]
Western Malaysia	n.a.	1.3	0.4	n.a.	32

[a]These measure the days of manual labour required annually for all crops and animals, assuming high western European standard requirements in all areas. Thus in Iloilo even transplanted rice is included at only 3 man-days per hectare, the same as wheat or barley in Aberdeen.
[b]Not including costs of own draught animals.

REFERENCES

Blaxter, K. L. (1978). Energy use in farming and its cost. In *Proceedings of the 32nd Oxford Farming Conference*, 1978.
Pimental, D. (1979). Energy and agriculture. In *Food, Climate and Man* (M. R. Biswas and A. K. Biswas, eds), John Wiley, New York, pp. 77–92.
Steinhart, J. S., and Steinhart, C. E. (1974). Summaries in *Energy Use in the US Food System, Science, NY*, **184.**

FURTHER READING

*Cleave, J. H. (1974). *African Farmers: Labour Use in the Development of Smallholder Agriculture*, Praegar, New York.
†Ghodake, R. D., Ryan, J. G., and Sarin, R. (1978). *Human labour in existing and prospective technologies of semi-arid Tropics of peninsular India*. Project report, International Crop Research Institute for the Semi-arid Tropics, Patancheru, AP, India.
Hudson, N. W. (1975). *Field Engineering for Agricultural Development*, Clarendon Press, Oxford.
*Makhijani, A., and Poole, A. (1975). *Energy and Agriculture in the Third World*, Ballinger Publishing, New York.
Pimental, D. (1979). Energy and agriculture. In *Food, Climate and Man* (M. R. Biswas and A. K. Biswas, eds), John Wiley, New York, pp. 77–92.
*Tschiersch, J. E. (1978). *Appropriate Mechanization for Small Farmers in Developing Countries*, Research Centre for International Agrarian Development, Heidelberg.

QUESTIONS AND EXERCISES

1. Give three examples of physical operations on farms that are still largely manual. Indicate why they are not mechanized.
2. How could uses of fossil fuels on and off farms, for the farming system you know best, be reduced without considerable reductions in outputs of crops and animals? What additional inputs, and what changes in practices, would be required?
3. In what ways can an able farmer secure higher than average efficiency in the use of his equipment, and reduce the repairs and depreciation on it relative to its output?

Part IV

The farm-economic sub-system

Chapter 10

Introduction

10.1 THE NEED FOR DECISIONS

The need for decisions is the reason for the conception of our third sub-system. The many differences that we noted, in Part II, in land use, cropping, animal numbers, use of biological inputs, total outputs and their composition, and, in Part III, in equipment and work methods – all these are the result of decisions by farmers. They were taken to suit different biological and work conditions and opportunities. And we noted how varied these conditions are between areas and even from site to site within areas. Tables 6.1 and 9.4 listed the most important variables. Unless decisions are well taken, with due regard to all these variables, good use cannot be made of the biological and work sub-systems of agriculture.

Decisions are also needed because of differences and changes in human population pressures on natural endowments, in non-agricultural employments and incomes, in transport and trade opportunities, in social organizations, and in personal and social values. These have been indicated in Chapter 1 and will be considered further in Part V. Such socioeconomic variables provide important criteria by which to judge values of agriculture outputs and inputs, including the annual flows from natural endowments. They therefore provide important 'choice criteria' which biological and engineering criteria alone cannot provide. Economics is the logic of choice when scarcities and values must be considered.

Examples of past decisions in different example farm groups are given in Tables 3.1–3.5, 9.5, 10.1, and 10.2. Examples of variations in decisions *within* particular areas are given in Chapter 8 and in Table 11.1. Examples of the effects of revisions of decisions are in Tables 1.9–1.11, Figures 11.1 and 11.2, and Section 14.4.2.

10.2 NATURE OF THE FARM-ECONOMIC SUB-SYSTEM

The farm-economic sub-system can be regarded as a most important decision-making (choosing) 'third tier', affected by and affecting the biological and work 'tiers' below it, and the socoeconomic 'tier' above. And within itself the farm-economic sub-system is affected by the wishes of the farm decision makers.

Close connections to the other 'tiers' must be recognized because decisions cannot be sound without knowledge from all four. Some thinking about agriculture still tends to be confined to biological and work matters, and to seek decisions without adequate knowledge from the socioeconomic system, and the disciplines of logic in the farm-economic. Thus ideals of 'good husbandry' arise along with doubts about any need for farm production economics. And similarly, when the logic of decision making is emphasized along with some socioeconomic knowledge, but not adequate biological or work knowledge, ideas of 'good husbandry' may be criticised. There should, however, be no such doubts and criticisms, because both knowledge and logic are essential to good decisions. We may usefully note that 'economics' and 'husbandry' are not very different words. '*Ekos*' from Greek and '*hus*' from Anglo-Saxon both meant 'household'. '*Nomos*' was derived from Greek ideas of logical laws. '*Buandi*' in Anglo-Saxon meant 'inhabiting', but came to imply doing so with thrift and success.

10.3   DEFINITIONS OF 'FARM'

Because the '*Ekos*' of 'farm-economic' should be distinguished from the wider ranging '*Ekos*' of modern socioeconomic systems, the word 'farm' should be defined.

'*Farm*' commonly means land (or water) and buildings used in the production of crops or animals or both. Sometimes it means to let out land-use or other rights. This is because 'farm' was derived from a low latin word, '*firma*', which meant both a fixed payment and a signature. 'Farm' came to mean first a 'lease' and then a 'tract of land held on lease'. 'Firma' came to mean the unit in the socioeconomic system that is legally responsible in buying, leasing, borrowing, selling, loaning, and other business affairs.

Farm firms make decisions about the production of crops, or animals (including fish), using land and water, or plant products (e.g. feedingstuffs, petroleum products) originally grown on these using photosynthesis. For convenience, their decision-making system we shall call simply 'farm-economic'.

The executive head of a farm firm is often called simply the 'farmer' or 'farm manager'.

The great majority of farm firms are individual heads of simple family households. But many other types of legal body are farm firms – partnerships, companies (corporations), extended families, collective farms, communities, and even local and central governments.

10.3.1   Additional goods and services

If in addition to crops and animals, farm firms produce other goods and services, the costs for these, and returns from them, may best be accounted (and statistically enumerated) separately. The decisions about them may, however, be interrelated with decisions about crops and animals, and should not be ignored when these farm-economic decisions are considered. (For example, in western

Europe and North America, many craft goods are made, and services provided to tourists and others seeking recreation. Many farmers do other non-farm work part-time. In the USSR many 'collective farms' have functions that elsewhere would be regarded as those of local government and engineering or building firms contracting with government.)

10.3.2 Relations to households

The relations of farm firms to farmers' families are especially important. Families consume farm outputs and supply management, labour, and often capital or credit to farm firms. Jointly with firms, they also commonly use buildings, motor vehicles, and other factors. Their wishes and advice are considered when farm firm decisions are made. For example, (a) improved farm mechanization can affect the attitudes of sons (Section 7.6); (b) improved equipment for some household work such as securing water (Section 7.5) can affect the labour available for field work in busy periods; (c) the decisions taken about changes in human work for the farm, as against other activities or leisure, often depend on individual and family values more than on market prices (Table 12.1)

In most regions, however, the decisions by individual firms and comparisons between farm firms are improved by accounting separately for family consumption of farm products and withdrawals of factors, and for labour and factors provided by families. The labour and management provided by farmers themselves and their wives have to be carefully defined to suit the purposes of particular comparisons or plans, although they may be difficult to value in money terms (Section 12.1.3).

10.3.3 Divided functions

Where responsibilities for the functions in crop and animal production are divided between legal bodies, further definition is needed of what is most conveniently for the purposes in hand regarded as the 'farm firm'. For example, where landowners are responsible for maintenance of buildings, contribute to some management policy decisions, and pay shares of fertilizer and other input costs, landowners' inputs and outputs may most conveniently be included along with tenants'. The farm firm can be accounted as a combination of landowner and tenant – that is on an 'owner–occupier' basis. Where landowners' functions are fewer and most land is tenanted, the farm firm can most conveniently be accounted on a 'tenant' basis, with rents as the cost of the use-rights bought from the landowner. Where (e.g. in the Sudan Gezira) services in land provision, irrigation, heavy machine work, pest control, and first-stage processing and marketing are provided centrally to tenant farmers, the farm firms considered may best be the tenants for some purposes, but, for other purposes, the entire central service firm and all the tenant firms can best be considered together.

Inevitably, misunderstandings and faulty conclusions result if care is not taken in choosing the most convenient definition of a 'farm', and stating it.

10.4   DEFINITION OF 'FARM-ECONOMIC'

The decision-making processes of farm firms, the knowledge of all kinds that is used by them in these processes, and the managerial skills and energies that carry the decisions into effect – these together are the farm-economic sub-system of agriculture.

'Farm-economic' is narrower in scope than 'agricultural economic', because this term is rightly used to include many aspects of the socioeconomic system as well as the 'farm-economic'.

'Economics' is frequently defined in ways that would include study only of those outputs and inputs that can be valued in money terms. But in the farm-economic sub-system, direct money measures are not available for important variables (e.g. improvements or deteriorations in soil structure; freedom from diseases and pests; the satisfactions from effective mechanization of physical tasks; the other non-material benefits that farmers obtain from their surroundings and work (Section 12.1.1). We should not, however, exclude from consideration in farm economics any variable that affects farm firm decisions, or is affected by them.

An older definition of 'economics' accommodates farm economics well: the study of 'human behaviour as a relationship between ends and scarce means that have alternative uses' (Robbins, 1932). We are concerned with the behaviour of farm firms. The desired ends are not simply money incomes and their stabilization (Table 12.1). The means in the biological and work sub-systems of agriculture are almost all scarce and can be used in alternative ways for alternative purposes.

10.5   SUMMARY OF FLOWS

A simple diagram cannot well represent the flows of knowledge required in farm-economic decision making. Nor can the processes by which the decisions are made and carried into effect be readily shown. But Figure 10.1 is useful in a showing important groups of components that have to be considered in a farm firm's economics (e.g. 'assets', 'purchases', 'production'). The flows between these groups during an accounting period are also shown. Although the components and flows are commonly valued in money, we should not ignore errors or omissions in such valuations.

10.5.1   Balance sheets

At the beginning and end of the accounting periods are statements of what the firm has for use (assets), what it owes (liabilities), and what it owns (net worth). In money terms, liabilities *plus* net worth, *equal* assets. These statements are each for one point in time, and the beginning statement for an accounting period is the end statement for the previous period. These statements have been likened to *still* photographs. The many transactions in the flows during an accounting period

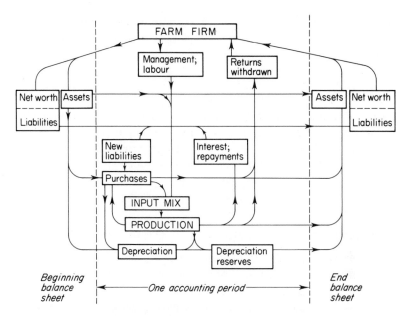

Figure 10.1   The farm-economic sub-system of agriculture: A simple diagram of flows.

which may be totalled to some operating statement require continuous recording as in a *movie* film. Commonly this recording is in accounts. But many of the world's millions of farmers have comparatively simple flows, and record them and their balance sheet positions only mentally.

10.5.2   Assets

Assets include climates, soils, plants, animals, and other components of the biological sub-system, and the machinery and other equipment of the work sub-system. Land and buildings are listed in statements of assets, and valued to include the natural endowments as well as any improvements by man.

To keep Figure 10.1 simple, it has been drawn for an owner-occupier. A tenant's assets would not include land but one of his purchases would be the use of land assets of the landowner. Rent in kind or cash would be paid. There would be other complications. (See Sections 15.2.2, 15.2.3, and 15.2.9.)

10.5.3   Flows from assets

From the assets at the beginning of the period four flows arise (see Figure 10.1): (i) the flow of assets themselves through to the end of the period (e.g. land, buildings, machines, breeding animals); (ii) the contribution of assets during the period to the 'input mix' that results in production (e.g. the biological flows made possible by land and its endowments, or by dairy herds, or the feeds or fertilizers in store that are used during the period); (iii) uses of assets to make purchases

142

(e.g. stocks of grain sold for cash, and cash used to buy fertilizers); (iv) depreciation of assets (e.g. of land through lack of weed control; of machines through wear and tear).

10.5.4 Purchases

Purchases are made possible by use of (i) assets; (ii) new borrowings (new liabilities); (iii) some proceeds of the year's production.

Purchases contribute to the 'input mix' (e.g. purchases of fertilizers). They may also help to offset depreciation (e.g. repairs or renewals of machinery) or to increase assets at the end of the year (e.g. purchases of additional equipment). Some purchases may form part of the withdrawals (e.g. fuel for house heating).

10.5.5 Depreciation

Depreciation (including depletion) of assets may be offset by (i) some proceeds of the year's production (e.g. growth of young animals to replace old in breeding herds); (ii) purchases (e.g. repairs or renewals). Depreciation that is not physically offset should be regarded as requiring deductions from the year's production to add to reserves that can be used to offset depreciation later. These reserves add to assets at the end of the period, but they also add the other side of the balance sheet where they may well be specially noted.

10.5.6 Flows from the farm firm

The farm firm provides (i) the *net worth* at the beginning year; (ii) *management* throughout; and commonly (iii) *labour* by the head of firm. Other family labour is, as already noted, best considered as 'purchased'.

10.5.7 Sources of production

The sources of production are essentially (i) the *management and labour* of the farm firm; (ii) flows from the land, plants, animals, and other *assets* in the biological sub-system. And these depend, as we have noted, partly on (iii) flows from the machinery and other equipment *assets* in the work sub-system. But both biological and work flows are affected by (iv) inputs (from *purchases* or beginning *assets*) or fertilizers, labour, and other items. Thus the whole 'mix' of factors of production has many components operating together. Many of them are flows rather than stocks, so that the timings of their uses are especially important (e.g. the use of the soil water and temperature conditions at seed time in temperate regions; Section 4.11). And management is very important in other ways (e.g. in decisions about which purchases to make, and when and how to use them).

10.5.8 Use of production flows

The outputs produced are used in various ways: (i) to pay for purchases and to offset depreciation or to add to reserves to be used to offset depreciation later; (ii) to repay debts (offset liabilities) and pay interest on debts; (iii) to pay taxes or provide reserves for later payment; (iv) to add to assets (e.g. increases in animal numbers or improvements in health or genotypes); (v) to provide cash or products, as returns withdrawn by farm firms.

10.5.9 Returns

The returns to farm firms include (i) these withdrawals from the proceeds of production; (ii) withdrawals from purchases made by the farm firm (e.g. fuel for motor vehicles); (iii) changes in net worth from the beginning to the end of the period.

All the components and flows in Figure 10.1 determine these changes in net worth. Poor flows from climates in particular years (poor weather), bad pest infestations, machinery breakdowns at critical times, and poor management – all these reduce production. Comparatively large withdrawals, or high taxes, reduce the flows to the end-of-period assets. New liabilities cause similar reductions if they do not, through well managed purchases, sufficiently increase production and end-of-period assets, or if they result in comparatively high interest charges not fully offset by increased production. Such considerations are of basic importance when farm firms plan to borrow, and banks and other bodies to lend to farmers (Section 18.2).

10.5.10 Valuations

The components and flows in Figure 10.1 are so diverse in kind that their use together in decision making is often possible only by expressing them in money terms. But we have already noted that some money valuations are difficult (Section 10.4 above). A general difficulty is that money valuations of assets at any one time should depend on what they can produce and on future prices; but assumptions about both production and prices are liable to error. Another difficulty is that changes in the value of a nation's money (e.g. pounds sterling; US dollars) cause changes in prices expressed in that money. Where inflation is rapid attempts are made to reduce the effects on operating accounts and balance sheets so that these can more accurately reflect real changes. In particular, attempts are made to distinguish between the total returns to firms that are due to (i) production relative to the 'input mix' – productivity in real terms; (ii) increases in asset values due to ownership (possession) (Section 12.1.3).

10.5.11 Subsidies and taxes

To keep Figure 10.1 as simple as possible, neither subsidies nor taxes are shown. Some subsidies may add directly or indirectly to market prices for products. Other subsidies reduce the costs to farm firms of purchases and the 'input mix'. Others are paid for production operations, and so reduce the net costs of the 'input mix' flows. Generally subsidies can best be seen as one of the flows from the

144

socioeconomic system *to* farm firms or to markets for factors (Figure 13.1). Taxes can be regarded as flows to government *from* the farm-economic sub-system, or related product or factor markets in the socioeconomic sytem. Income taxes are on net flows from farm firm production to returns withdrawn and end-of-period 'assets'. But other types of tax raise transport costs and factor prices, and reduce product prices. Other taxes are based on the total values of assets (e.g. local property taxes).

In addition to subsidies and taxes, other components of the socioeconomic system affect farm firms and these will also be considered in Part V.

10.6   DATA FOR EXAMPLE GROUPS

Because biological and work flows can be summarized in money terms, some data have been presented in Parts II and III (e.g. Tables 3.4, 3.5, and 9.5). Balance sheet data are summarized in Table 10.1 and show big contrasts in total assets per farm and in the make-up of these assets (e.g. as between the Iloilo and southwestern Louisiana groups, and between the Norfolk and Illinois groups).

Annual 'input mixes' are partly summarized in Table 10.2. The annual value of inputs per hectare of total farm area varied from US$53 in Aurepalle, southern India, to US$1324 in Norfolk. But Table 10.2 does not include any value for inputs from (i) land and buildings; (ii) use of capital; (iii) management; (iv) risk and uncertainty bearing; (v) services paid for by taxes. The returns on these inputs are conveniently regarded, along with returns on manual labour, as shares in the *total 'net value added'* by the farm firm's operations (Table 12.2). Within the 'input mixes' measured in Table 10.2, the relative values of 'biological' inputs

Table 10.1   Balance sheet items: average farm, example groups, end of year 1978–9.

Groups	Short, medium term	Long term	Liquid	Crops and animals	Machinery	Land and buildings	Total
				US$ per US$100 of total assets			US$'000
Aberdeen	14	3	3	28	11	58	382
Norfolk	4	3	2	17	12	69	759
Cortland, NY	12	19	3	27	14	56	629
Illinois	5	8	+	7	7	86	1252
Saskatchewan	+	8[a]	9	15	15	61	278
Louisiana	n.a.	n.a.	2	5	14	79	919
Iloilo	n.a.	n.a.	1	7	1	91	7
Aurepalle	n.a.	b	2	10	10	78	5
Western Malaysia	n.a.	n.a.	3	73	2	22	7

[a]Total liabilities.
[b]Total end-of-year liabilities were 1.4 per cent of total assets.

Table 10.2   Variable and 'fixed' costs: average farm, example groups, 1978–9.

Groups	Total per hectare of Total area[a]	Biological	Labour[b]	Machinery use	Other	Total
	US$	*US$ per US$100 gross output*				
Aberdeen	832	29	20	20	5	74
Norfolk	1324	28	17	22	6	73
Cortland, NY	853	37	12	22	7	78
Illinois	407	18	10	23	5	56
Saskatchewan	146	14	13	26	6	59
Louisiana	301	21	11	26	9	67
Iloilo	503	7	41	4	1	53
Aurepalle	53	18	13	6	0	37
Western Malaysia	253	7	41	4	7	58

[a]Excluding rough grazings and woods.
[b]Including all manual labour.

(such as fertilizers, pesticides, seeds, veterinary services, feeds) varied widely. The ratio of manual labour to machinery use also differed greatly (e.g. between Aberdeen and Cortland, NY; south-western Louisiana and Iloilo; central Norfolk and central Illinois). (See also Table 9.5.)

REFERENCE

Robbins, L. (1932). *The Nature and Significance of Economic Science*, Macmillan, London.

FURTHER READING

Abbott, J. C., and Makeham, J. P. (1979). *Agricultural Economics and Marketing in the Tropics*, Longman, London.
Bannock, G., Baxter, R. E., and Reese, R. (1972). *Dictionary of Economics*, Penguin, Harmondsworth.
Cramer, G. L., and Jensen, C. W. (1979). *Agricultural Economics and Agribusiness: An Introduction*, John Wiley, New York.
Halcrow, H. G. (1980). *Economics of Agriculture*, McGraw-Hill, New York.
*Hardaker, J. B., Lewis, J. N., and McFarlane, G. C. (1970). *Farm Management and Agricultural Economics: An Introduction*, Halstead Press, Sydney.
Hill, N. B. (1979). *Introduction to Economics for Students of Agriculture*, Pergamon Press, Oxford.
†Jensen, H. R. (1977). Farm management and production economics, 1946–70. In *Survey of Agricultural Economics Literature*, Vol. 1 (L. R. Martin, ed.), North Central Publishing (for the American Agricultural Economics Association), St Paul, Minn.
Sturrock, F. (1971). *Farm Accounting and Management*, 6th edn, Pitman, London.

QUESTIONS AND EXERCISES

1. Briefly describe the nature of the farm-economic sub-system of agriculture.
2. What is the commonest type of farm firm in the region you know best? What other types also provide or use farm land? What types (if any) are increasingly important?
3. For one particular crop, list the components in the 'input mix' for a year and list the 'outputs'. In both lists include a second section with items that are especially difficult to value in money terms.
4. Find a recent annual report on the incomes and expenditures of farm firms. Compare the data for a group of farms with the data in Tables 3.1–3.5 and 10.1.

Chapter 11

The decisions made

11.1 TYPES OF DECISION

Before considering decision-making methods in Chapter 12, we can here usefully examine further the types of decision made by farm firms. We can relate them to Figures 10.1, 3.1 and 7.1, and to the socioeconomic system (Figure 13.1). The importance of the results of farm firm decisions will thus become clearer, and some of the problems to be faced in Chapter 12 will be better understood.

Decisions can be considered as answers to questions, and classified under the following general questions. The headings that economists use for the classes are in brackets.

(1)   What to produce? (Choice of products *or* composition of output)
(2)   How much to produce? (Quantum *or* volume of output)
(3)   How to produce? (Factor proportions *or* combination of factors in the 'input mix')
(4)   How much to invest? (Investment)
(5)   How much to save and borrow? (Savings, and use of credit to obtain loans)
(6)   What to hire or lease? (Renting, hiring, leasing)
(7)   Where and how to buy? (Factor purchasing)
(8)   Where and how to sell (Product marketing)

Each of these eight general questions can be regarded as including questions about the carrying into effect of the detailed answers (execution), and continual assessment of results to ensure control, and to provide knowledge on which to base revised answers to suit the changing conditions (monitoring and feedback). Each general question should be regarded as including the words 'and when', because timing is important within each group of decisions.

An important further general question about the farm firm's activities as a whole is

(9)   When to start, and when to stop?

Some questions have answers that have long-term consequences (e.g. those

147

related to (9), and many related to (1) and (3)). So in practice individual firms are not making decisions continually over the whole range of detailed questions. But we should understand the whole range because it affects new firms, and eventually all firms, and because logically all answers should be compatible.

The relations amongst questions (1)–(8) are close (e.g. of (3) to (2); (6) to (5); (5) and (6) to (4); (8) to (1)). Because they are so close between (2) and (1), these are considered together here.

11.2   CHOICE OF PRODUCTS AND VOLUME OF OUTPUT

11.2.1   Where choices are limited

Wherever climate, soils, and other biological conditions would not, without *excessive* investments of labour and equipment use, permit more than a limited change from natural vegetation, choice of products is obviously closely restricted (e.g. in the mountains of northern Scotland; in wide areas of Africa, with low rainfalls, or tsetse flies and sleeping sickness).

Similarly, if the work sub-system imposes severe constraints, choice of products is limited (e.g. in some sparsely populated areas of Brazil). But supplies of labour and capital, demands for foodstuffs, provision of transport, research to find ways to improve some soils at low cost – all these and other socioeconomic variables may change and make practicable changes from the more natural vegetative cover to cultivars.

Some biological conditions favour particular crop species that are elsewhere especially difficult to grow (e.g. rubber in western Malaysia; quality tea in some highland tropical areas; sisal in some lowland). In Scotland, 90 per cent of the production of raspberries is concentrated in certain localities in only two counties. In the USA, almost 90 per cent of all hop production is in only two states. In each of the localities of concentration, the especially favoured crop species is now the obvious choice as the major crop. Management and labour skills and equipment, and market arrangements for factors and products, have been developed and tend further to favour the major crop.

11.2.2   Where choices are not so limited

Choices of product are most commonly made in biological circumstances that do not so much favour only one major product. Decisions therefore become more intricate. Chapter 12 introduces logical decision-making methods. But here the reasons for intricacy are reviewed by (i) considering briefly the general advantages of producing several products (diversifying), and, on the other hand, of producing few (specializing); (ii) considering how changes in conditions, knowledge, and goals require detailed changes in decisions.

The advantages of diversifying   The *biological* reasons may perhaps best be recalled from Figure 3.1 and Chapter 4 and 5. Variations of soils and micro-climates within farms can be better suited if several crop species are grown. Sequences of different plants, and mixtures of crop species tend to make better use of soil and subsoil minerals, nitrogen fixed by legumes and other residues or by-products of particular species (e.g. straw and roots of cereals, soil structure improvements of rotational grasses). Mixed cropping may make better use of climates. Rotation of crops helps to control biotic factors, and soil erosion in some circumstances. Animals provide outlets for crops that are suited to particular climates and soils but whose outputs are not, or not entirely, marketable as human food (e.g. maize, barley, hay, silage). In many areas both ruminant and non-ruminant animals are needed if the mixed output of concentrated and roughage feedingstuffs is to be well used. Some animal outputs are by-products (e.g. male calves on dairy farms, wool from lambs for meat). Animal faeces and urine (dung) serve as soil fertilizers.

The *work* system also explains important reasons for diversifying. The work inputs for most individual products are seasonal. The potential flow of labour from workers who are on farms throughout the year is more regular, week by week. Therefore to make best use of the whole potential annual flow, various products can be chosen so that the seasonal inputs for them have together a more regular total labour requirement. (For example, the requirements for barley, grass, and cattle together, in our Aberdeen area, are much less seasonal than those for barley alone.) Another work reason for diversifying in some areas is the traditional division of labour. Some products are regarded as needing women's work more than men's. Some tasks for some products are for children.

Other *farm-economic* and *socioeconomic* reasons for diversity are important. Market prices vary over time (Section 14.3.1). Therefore, the ratio of one product price to another product price varies. By producing more than one product the total value of 'production' is usually more *stable*. Also firms that are equipped to produce more than one product may be more *flexible* in adjusting production plans to longer term changes in price ratios. Moreover, the potential income to farm firms from different products varies *seasonally* with the biological timing of harvests, slaughter dates, etc. Having more than one product may be desirable to make the flow of income more regular (Section 18.2). Within areas, *young and old* farm firms tend to differ in the ratio between liabilities and net worth, and in total assets, and therefore to have different choices of product.

To secure approved supplies of foodstuff for *farm family consumption* each week production is often diversified (e.g. to include vegetables, eggs, and milk).

Where traditionally wives are partners in farm firms but their withdrawals are the proceeds of particular products, these may be favoured among other products (e.g. until the 1950s in Aberdeenshire, wives were concerned with much egg production).

The advantages of specializing Specialization helps to foster the full development of special knowledge and skills, and so greater biological, work, and marketing efficiencies. Biological, work, and farm-economic efficiencies may be improved by having more specialized equipment (e.g. power sprayers, combine harvesters). Purchase of equipment, chemicals, feeds, and other factors may be at lower costs.

Various management functions are simplified, although others may be made more difficult (e.g. disease and pest control).

11.2.3 Diversity and specialization in example groups

Of the example groups in particular areas the two nearest to having only one main product are Cortland, NY, and western Malaysia (see Table 3.5). The next most specialized are the wheat farms of Saskatchewan and the beef farms of Aberdeenshire. The least specialized are in central Norfolk and Aurepalle. Even in the Cortland area, with milk as the major product, choices are made about the proportion of the heifer calves to be reared. These determine sales of young calves, culled cows for meat, and heifers and cows for milk production elsewhere. There are also some sales of crops (Table 3.5).

In all the example groups, development of the socioeconomic system has affected the decisions about choices of products. The most important components have been (i) markets and prices for farm products, including transport and storage facilities; (ii) supplies and prices in factor markets (e.g. for fertilizers, seeds, hired labour); (iii) supplies of loans, and interest rates; (iv) taxes and subsidies; (v) government policies on research and extension education; (vi) social values and goals.

These are considered in Chapter 14–21, but here we should note particularly the importance, variety, and changeability of all these components as affecting choice of products.

11.2.4 Decisions about choice of products, change over time

Farm firms have to repeat the decision-making process in whole or in part, and many more often than yearly. This is because (i) as just noted, components of the socioeconomic system are changeable (e.g. grain prices can rise and beef prices fall); (ii) biological conditions change between years as well as between seasons (e.g. poor weather may reduce an area's feedingstuffs supply at a time when prices are high so that cattle production plans must be altered); (iii) knowledge increases about relationships in and between the biological, work, and socioeconomic systems (e.g. from results of cultivar trials and pest control studies); (iv) the net worths and liabilities of individual firms change (e.g. as young men progress from heavy borrowings relative to their net worths and perhaps low total assets); (v) values and goals of individual firms change (e.g. aversions may become greater to high year to year fluctuations in money incomes).

Figures 11.1 and 11.2 give further examples of how choice of products changes over time. Acreages of annual crops in temperate regions can of course be change only once a year. The irregularity of the changes indicates how many and wide are the variations affecting the decisions (Figure 11.1). Some choices in livestock

151

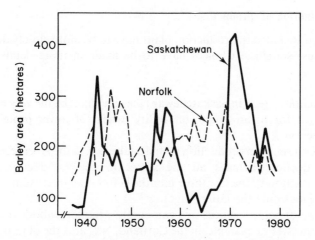

Figure 11.1   Barley area per 1000 ha wheat in Saskatchewan and per 100 ha wheat in Norfolk, England, 1938–78.

production are less frequent than yearly but some are more frequent. Important 'cycles' of more regular changes can occur, but most changes are due to irregular variations (Figure 11.2).

In considering changes, year to year and even over longer periods, we should of course recognize that many fundamental determinants of choice of product do not change. Major characteristics of farm production plans have remained unaltered (e.g. the ratio of grassland to grain crops in the Aberdeen area – except in war-time; the dependence of Saskatchewan on spring sown grain and of western Malaysia on tree crops).

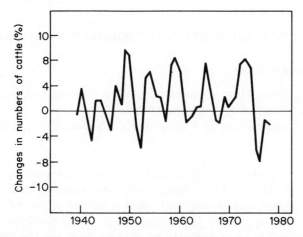

Figure 11.2   Year to year changes in numbers of cattle under 2 years old, England and Wales from 1938–9 to 1977–8.

11.2.5 Definition of 'product'

The products included in production plans have to be more closely defined than those in most statistics. Choices have to be made in more detail about the following:

(a) *Qualities* (e.g. in Saskatchewan) the aim might be the very high quality wheat for bread flours, or 'utility' wheat of lower protein content; (Section 3.1.2)

(b) *Which stages* in production to undertake (e.g. in the Aberdeen area some production includes all the stages in beef production from calf production to the last three months before slaughter: some production includes only the latter stage)

(c) *Time of sale* Choices are commonly necessary about the timing of outputs. (For example, in the Cortland, NY, area the objective might be level monthly outputs of milk, or more milk in the months when production is cheapest. Time of sale may also be varied if storage equipment and finance are available to farm firms. In the Norfolk, Illinois, and Saskatchewan areas, grain may be stored for variable periods after the harvest times).

11.2.6 Differences between farms

Choice of products commonly differs considerably *within* areas. Farm by farm differences in many components of the biological, work, and farm-economic subsystems are often quite substantial (Sections 4.1, 4.9, and 9.3). And virtually always, differences in values and goals, assets and net worth, knowledge, and enterprise of the farm firms cause some differences in choice of products at any one time and in changes over the years (Chapter 12).

11.3 FACTOR PROPORTIONS (COMBINATION OF FACTORS)

11.3.1 Classes of factor

The decisions about 'How to produce?' relate to the composition of the 'input mix', including the timing of the use of factors.

Figure 10.1 and Section 10.5 show that the 'input mix' is made up of the following:

(a) Annual flows from assets that are either (i) used up during the accounting period (e.g. feedingstuffs, seed, fattening cattle sold) or (ii) not so used up but remain as assets at the end of the period, although perhaps depreciated or depleted (e.g. land, buildings, machinery, plantations); occasionally some of these assets may be sold

(b)   Purchases of fertilizers, casual labour, and other inputs that are used during the year rather than stored

(c)   The management and labour of the farm firm and regular year-round labour.

Classifying inputs in this way is useful because decisions about class (a) (ii) have mainly medium and long term consequences and those about class (a) (i) and (b) have shorter term consequences. Most of the class (a) (ii) assets are commonly called 'lumpy' because they are comparatively large. Class (a) (i) and (b) inputs are 'divisible' into small units, and so the amounts used can be closely adjusted. Class (c) inputs are flows of various kinds (skills, managerial energies, and labour) – commonly from the farm household head and his wife (if she is regarded as joint owner of the firm). They also include regular (year-round) labour. The ratios between these (c) inputs and all other inputs are important, as we shall see when considering sizes of farm firm businesses (Section 12.4.6). The appropriate timings of class (c) inputs are also important.

11.3.2   Shorter term output–input relations

Decisions about fertilizers and other inputs used up within an accounting period (classes (a) (i) and (b)) are in character somewhat simpler than other 'input mix' decisions. For example, nitrogen fertilizer applied in year 1 is regarded as completely used in that year. The additional grain or other crops produced in response is forecast only for year 1, although there may be also additional crop residues to improve soil fertility for year 2. Phosphate and potash fertilizers and lime may be regarded as having longer-term effects, but simple shares of these are commonly charged to year 1. Likewise responses to feeds used in year 1 are all regarded as arising as milk, or other animal outputs, in that year (including any improvements in the end-of-year assets in animals, although in practice these may be difficult to value reliably).

Figures 4.2 and 11.3–11.5 give examples of output–input relations. Those for sugar-beet responses to nitrogen in two localities in the east of England call attention to how important climatic and soil factors can be, when decisions about nitrogen are taken. Similarly the responses to nitrogen of potatoes in the Aberdeen area (Figure 11.3) depend on how much potash is applied, so that some 'balance' between nitrogen and potash has to be decided (Section 4.9.7). The actual responses of milk production to additional inputs of concentrated feedingstuffs are determined by the many variables considered in Chapter 5 and Section 8.3. Figure 11.4 is based on experimental work (Knoblauch and others, 1978); in practice responses in many different herds would be less linear because of genetic and managerial restraints. Additional pig-meat production can result from increased feed inputs but the output–input relations are complicated by differences in carcass composition (Figure 11.5). The output of fat responds differently from the output of lean meat, and both energy and protein intake rates

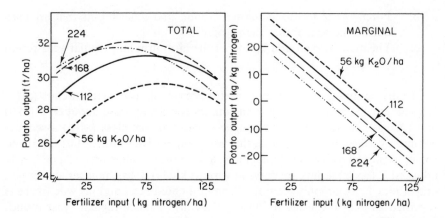

Figure 11.3 Outputs of potatoes related to inputs of nitrogen fertilizers, Aberdeen area. The responses of total product differed at different levels of potash (K_2O) inputs. The marginal products measured the additional weight of potatoes (in kilograms) per additional kilogram of nitrogen input. They diminished with higher levels of nitrogen and K_2O inputs. (Redrawn from Barlow, 1964.)

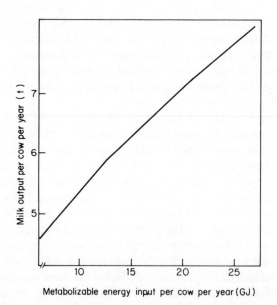

Figure 11.4 Outputs of milk related to inputs of metabolizable energy in concentrated feed. Cortland, NY, group planning of corn silage-based diets, 1978. This output–input relationship was estimated from energy 'requirements' in experimental animals.

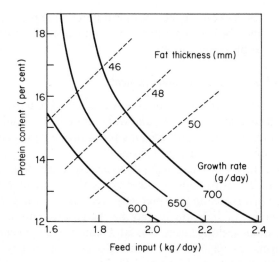

Figure 11.5 Output of pigs (growth rate and its quality as measured by shoulder fat thickness) related to inputs of feed (dry weight and crude protein content).

have to be considered as well as the ages and weights, genotypes, and environments of the pigs (Crabtree, 1976).

11.3.3 Timing as affected by management and equipment

The responses of outputs to better *timing* of inputs (including attentions to animals) are more difficult to take into account. These responses are not usually directly measured in experiments or surveys although some are (e.g. the effects of late sowing; Figure 4.4).

Better timing generally depends both on (i) better skills and energies in management and labour and (ii) greater capacity for work, and greater reliability of machinery and other equipment, in relation to the tasks to be carried out. We are led therefore into decisions about factors of the types (a)(ii) and (c).

11.3.4 Input flows from 'lumpy' assets

Many of the flows into input mixes are from major assets such as land, with its natural endowments of climate and soils and its man-made improvements (e.g. drainage and irrigation systems, freedom from weeds, favourable soil structure, soil conservation layouts; Section 4.9). These endowments and improvements provide year by year into the future a sequence of input flows. If further improvements are to be made (or if there are to be erosions or depletions), the decisions should be based on consideration of output–input relations over much more than one year. This is also true of decisions about other 'lumpy' assets such as tractors, combine harvesters, and other machines. Orchards and plantations similarly provide year by year flows of inputs and alternative production plans can have different results because they include different intentions about types of cultivar planted, ages before replanting, pest and disease control measures, pruning policies, tapping systems, etc.

11.3.5 Ratios of other inputs to management

A very important set of decisions is about how much land of particular qualities, machinery, livestock, and other assets, and what input flows, should be managed by particular farm firms. These are decisions about *sizes of businesses* – about the ratio of class (a) and (b) to class (c) factors.

Of these class (c) factors, management flows are especially difficult to measure. What individuals, or partnerships, or other types of firm are able and willing to supply over a run of years cannot be precisely predicted. In whole economies the abilities and energies of individuals may be increasing, and so too may the number of individuals seeking farms. But land tenure arrangements may favour existing farm firms, as against new firms. Government policy decisions are therefore demanded (Sections 15.2.1 and 15.2.5).

11.3.6 Examples of differences and changes in factor proportions

The average farms in our example groups differed in their input mixes (Tables 3.4; 9.5, 10.2; 12.2). The basic causes included all those that determined choices of products and quanta of production. Also the supplies and prices of factors had major influences. In different areas, there were different ratios between the costs of particular input flows. (For example, in the Aberdeen area, costs per hour of labour were higher and of tractor use much lower than in Aurepalle. So the labour: tractor use factor price ratio favoured use of less labour and more machinery in the Aberdeen area (Table 10.2).) The wide differences between North American input mixes and those in Asia and Africa are especially noteworthy.

Within our example groups the differences in input mixes were also wide (Table 11.1). They reflected many biological, work, and farm-economic differences between farms, including differences in managerial skills and decisions of the farm firms.

Changes in input mixes over time are made in response to (i) changing knowledge of biological and work relationships; (ii) changing prices of factors and products; (iii) changing goals; (iv) other changes (Sections 12.1, 12.2, and 12.4). Thus in Saskatchewan during the 1970s the inputs of hired labour, machinery use, and fertilizers had contrasting trends and fluctuations (Figure 11.6). Even in Iloilo substantial changes were made as knowledge was gained of new cultivars, multiple cropping possibilities, and wet seeding (Barlow *et al.*, 1983).

11.4 INVESTMENT DECISIONS

Figure 10.1 shows that the sources of investment are *production* (less taxes and servicing of loans), new *liabilities*, and reduced *withdrawals* (savings).

Table 11.1   Examples of variations between farm firms due to differences in their input mixes.

Farm group and variable	Coefficient of variation[a]
Aberdeen, 1977–8 and 1979–80 averages	
Land area per farmer	61
Tenant's capital per hectare	31
Variable costs per hectare	44
Gross output per hectare	31
Land area per £1000 work costs[d]	27
Management and investment income per hectare	79
Central Norfolk, 1978	
Land area per £1000 manual labour costs	63
Land area per £1000 machinery use costs	42
Gross margin per hectare[b]	48
Dairy farms, New York State, USA, 1979	
Maize silage yield per hectare	31
Milk sold per cow	17
Milk sold per man-equivalent	33
Iloilo, Philippines, 1978	
Production per hectare[c]	35

Sources: See Table 3.1.

[a]For each variable listed, this measures the variation round its mean. On approximately one-sixth of the firms, the variable exceeded the mean $\times [1 + (\frac{\text{coefficient}}{100})]$ and, on approximately one-sixth, was less than the mean $\times [1 + (\frac{\text{coefficient}}{100})]$.

[b]Gross output less biological and other variable costs (excluding labour and equipment use).

[c]Net cash surplus plus household expenses *minus* non-farm income.

[d]Manual labour and machinery use.

Particular assets may of course be increased *within the total* (e.g. selling an old small combine grain harvester and breeding ewes and using the cash proceeds to buy a big new grain harvester).

Decisions on investments depend on assessments of (a) how productive the additional assets are going to be, and when, as against (b) the value to the firm of the potential withdrawals that might not be taken, and/or (c) the costs of additional liabilities. These costs include interest charges and potential loss of flexibility in future decision making if lenders require repayment at times when production is low and the firm wishes to avoid reductions in assets, and/or withdrawals (Section 18.2). In practice (a), (b), and (c) are all difficult to assess. But they are of fundamental importance as determinants of change in agriculture because this depends on total asset flows into 'input mixes' (Figure 10.1). When particular farm firms wish to increase their liabilities, but cannot because limits are set by banks and other lenders, capital is said to be *involuntarily rationed*. Liabilities are *voluntarily rationed* when the firms could borrow more, but do not wish to do so (Sections 18.B and 18.C).

158

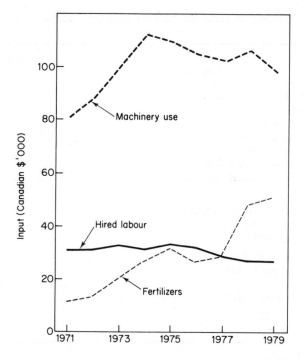

Figure 11.6   Major types of input in Saskatchewan, 1971–9, in Canadian dollars of 1971
value.

11.5   SAVINGS AND BORROWINGS

Decisions about *withdrawals* by firms and about *liabilities* depend on production
levels and investment decisions (Figure 10.1). For any particular pair of
production and investment levels, decisions have to be made about the ratio of
borrowings to savings. Savings within farm households of money actually
withdrawn (or obtained from non-farm sources) and invested outside the farm
firm are not considered here. Savings are, for simplicity, regarded here as
reductions in *withdrawals* from the farm firm itself. The results of the decisions
are reflected in year to year changes in the ratio of liabilities to net worth in
balanced sheets. In many areas, this ratio is kept comparatively low so as to avoid
interest charges and financial difficulties in periods of poor production or low
product prices. Thus liabilities were less than 20 per cent of total assets in several
of our example groups (Table 10.1). But in some circumstances, where
production is low (e.g. in Iloilo and Aurepalle) withdrawals by farm firms as
income to the farmers' households are already so low that they permit
consumption of little more than essentials. Additional assets and purchases of
fertilizer, seed, agrochemicals, and other inputs during growing seasons require
substantial new liabilities (Section 17.2.4).

Chapter 18 is concerned with the provision of loans to farm firms so that (i) their total *assets* may be appropriate to their production possibilities and (ii) their total *liabilities* may be appropriate to their *net* worths, production possibilities, and attitudes to possible fluctuations in the value of *production* and therefore in *withdrawals*, total *assets*, and loan repayments.

11.6   LEASING (RENTING) AND HIRING

The flows from some types of asset may be secured without having full ownership of them. Land-use rights are an example because land may be leased for one season, or year, or for a span of years. Rights to use related buildings and other equipment of the land are commonly included. Even dairy herds may be leased in some areas.

Some equipment may be leased for very short periods (machinery hire). The carrying out of some specific tasks by expensive equipment and its drivers may be paid for (e.g. contract work in combine harvesting of grain). In these ways, the input flows from lumpy assets are divided amongst more farm firms (Section 17.2.3). Some loss of control of the timing of work commonly occurs.

Leasing and hiring reduce the assets that firms have to finance by net worth or liabilities, but net production may be reduced by little more than the rental or hire charges. The assets that are financed may overall be substantially more productive. This is a reason why tenant farmers commonly have higher net return rates on their total assets than do owner occupiers. But this may not be true where land is very scarce, those seeking tenures are many, and tenures are insecure. Or where machinery hiring or contract work leads to faulty timing.

11.7   FACTOR PURCHASING

The factors that are purchased so as to secure the 'input mixes' decided on are, in well developed economies, usually available from a number of sources that compete with each other (Section 17.3). Decisions have therefore to be taken by farm firms about when and from whom to purchase.

In many localities in poorer countries, the links to well developed factor markets are less satisfactory. In such circumstances, the scope for farm firm decisions may be very limited in the short term. But some decisions may be desirable about methods of increasing competition by strengthening links with more central markets (Section 17.2).

In well developed economies, competition may also be reduced, so that farm firms have special decisions to make. Some firms marketing agricultural produce (e.g. eggs) provide farm firms with factors (e.g. chicks and feed) on credit, along with technical advice, and require produce to be sold to them. They thus use legal contracts with farmers to integrate factor marketing, production, and product marketing. If the integrating firms are few, their pricing behaviour may tend to raise prices of inputs relative to prices of outputs. When firms themselves expand to undertake all the functions including those of farm firms, they are said to be 'vertically integrated'.

11.8 PRODUCT MARKETING

The nature of farm firm decisions about product marketing also depends on how fully produce markets have been developed (Chapter 14). In all localities, some decisions have to be taken by farm firms about *when* to sell and in *what forms* exactly – quality, packaging, sizes of lots, etc. These may be regarded also as detailed decisions about choice of products (Section 11.2) above, but their dependence on detailed information from produce markets is especially close. In localities with competition, 'where' and 'to whom to sell' have to be decided. In localities with one buyer (monopsony), decisions have to be taken about 'how' and 'when' to bargain and about 'whether to organize' and strengthen links with more central markets (Section 14.4.4).

Because prices in product markets fluctuate and cannot be reliably predicted, some farm firms wish to reduce their uncertainties by accepting guaranteed prices for future deliveries. Others wish to receive income early, and even sell crops months before harvests. Governments may also give some forward price guarantees. But whatever are the various possibilities of reducing uncertainties about produce markets, farm firms have decisions to make and require information on which to base them.

11.9 BEGINNING AND ENDING A FARM FIRM

For many farm firms, the timing of their beginning was a major determinant of how good or bad their results were. This was because (a) the relation of the prices of their products became more favourable or unfavourable in relation to the values of their assets when they started and to their initial liabilities, and/or (b) they operated in a period when produce prices became high or low relative to the prices of factors they bought (e.g. feeds, hired labour, loans). Market and monetary changes in the socioeconomic system determine such relationships, not an individual firm's decisions (Table 14.2 and Section 19.4.4).

The starting of a farm firm by the purchase of 'lumpy' assets commits 'liquid' assets (such as money owned or money borrowed) to 'non-liquid' forms. These may rise or fall in value, and may or may not be sufficiently productive. Reconverting to 'liquid' form would incur disturbances of production, and costs. The timing of investment in 'lumpy' assets is therefore of major importance.

Many farm firms start because assets are inherited, and are already largely in 'non-liquid' forms. Decisions may be necessary about their valuation, and about what liabilities to assume (e.g. so that others sharing the inheritance can have 'liquid' assets). Such decisions involve decisions about whether or not to start a farm firm. Sons who inherit some of their fathers' assets and continue to farm where their fathers farmed, do decide to start new firms.

Decisions to end the operations of a firm are often difficult because farm firms are largely financed by individuals or families. Transfer of assets to succeeding firms during the lifetimes of particular farmers may be difficult for personal reasons, and because satisfactory provisions for the farmers in retirement are not made. Many farmers continue in control long past the ages of retirement from other occupations. To avoid this, and to avoid excessive taxation of capital transfers, decisions are increasingly taken in Europe and North America about (i) the formation of small corporations so that the deaths of individuals do not determine the lengths of lives of firms, and (ii) father–son agreements including fathers' retirement plans.

11.9.1 Part-time farming

Some farm firms occupy their owners for only part of their working time. Their farms may indeed provide little in addition to homes. But they may have produced more in the past, and could do again if choice of product and 'input mixes' were altered, perhaps by some new firm. In part-time farming, therefore, firms can slow down rather than stop. And new firms can start more readily than full-time farm firms.

11.10 VARIATIONS IN THE RESULTS OF DECISIONS

We have noted that decisions of every type vary from area to area, from firm to firm within areas, and also over time. Their results therefore also vary. Examples have been given for some types of decision.

The natural and human differences that determine all these variations also determine the scope for improved decision making for the future. It is this that will determine the directions and paces of development. Some improvements can begin in the socioeconomic system (Chapters 13–20). Other improvements can begin in the farm-economic system itself. But *all* improvements require new decisions in the farm-economic system, and better understanding of the essentials of rational decision making by farm firms (Chapter 12).

REFERENCES

Barlow, C. (1964). Input–output relationships and economic choice. PhD thesis, University of Aberdeen.
*Barlow, C., Jayasuriya, S., and Price, E. C. (1983). *Evaluating Technology for New Farming Systems: Case Studies from Philippine Rice Farming*, International Rice Research Institute, Los Banos, Philippines.
*Crabtree, J. R. (1976). Profit-maximising diets for growing pigs. In *Feed Composition, Animal Nutrient Requirements and Computerization of Diets* (P. V. Fonnesbeck *et al.*, eds), Utah Agricultural Experiment Station, Logan, Utah.
Knoblauch, W. A., Milligan, R. A., and Woodell, M. L. (1978). *An economic analysis of New York dairy farm enterprises*. AE Res. 78–1, Department of Agricultural Economics, Cornell University, Ithaca, NY.

162

FURTHER READING

As for Chapter 10, and the following:

Dalton, G. E. (1982). *Managing Agricultural Systems*, Applied Science Publishers, Barking, Essex.
Giles, A. K., and Stansfield, J. M. (1980). *The Farmer as Manager*, Allen and Unwin, Hemel Hempstead.
Upton, M., and Anthonio, Q. B. O. (1979). *Farming as a Business*, 2nd edn., Oxford University Press, London.

QUESTIONS AND EXERCISES

1. List and briefly describe the decisions that determine changes in the outputs of two important products from farms in the area or region you know best.
2. During the last 10 years, what have been the most obvious changes in the 'input mix' used on farms in the area or region you know best?
3. Give five examples of investments by farm firms in 'lumpy' assets, and indicate the common lengths of life of the assets.
4. Farmers' net returns vary widely even within local areas. List eight types of decision that may cause such variations.

Chapter 12

Essentials in rational decision making

Decisions in the farm-economic sub-system cannot be well made unless farm firms have (a) clear ideas of their goals (objectives), and what progress they make towards them; (b) knowledge of all aspects of the biological, work, and socioeconomic systems as affecting their own farm-economic system's operations and opportunities; (c) logical procedures. This chapter considers these three groups of essential in Sections 12.1, 12.2, and 12.3, and then introduces the range of methods that have been developed to help firms to reach rational decisions (Section 12.4).

Once made, decisions should be carried into effect. And the results should be assessed to provide 'feedback', as part of the knowledge on which further decisions are made. This chapter therefore also notes the importance of the monitoring of operations and control of results (Section 12.5), and the man-management and organization in farm firms (Section 12.6).

12.1 GOALS AND VALUES

12.1.1 A list

Almost all farm firms have many goals. They arise from their basic value judgements. Thus a higher average net annual income is a goal arising from the values placed on higher levels of consumption or larger additions to net worth. The higher income would be '*instrumental*' in securing these ends. A lower year to year variation in net income flows would be instrumental in providing greater 'peace of mind'. Fishing, shooting, and other leisure pursuits would be recreational, and within limits would help to raise incomes and make these more secure. Beyond these limits management and labour inputs would be smaller, and incomes would be lower and less secure, but the additional leisure might still itself yield *intrinsic* and *aesthetic* values to individuals. Breeding prize-winning animals and other biological achievements, or following a particular traditional system of farming, have high '*expressive*' values to some farmers, because thereby they can express their personal abilities and aptitudes.

Many goals arise from *social* values in improving relations of farm firms to farm families. These social values are commonly closely interconnected to instrumental values, and related to timing. For example, higher annual net income flows can be instrumental in securing the following: (a) higher levels of family consumption in the near future (including durable consumer goods to ease and speed housework); or (b) higher net worths as a basis for (i) more mechanization to ease the family's agricultural tasks; (ii) investments that will increase the firm's income later; (iii) the establishment of a son on another farm in the future; (iv) earlier retirement of the present farmer with more adequate and secure income, and handover to a son. Therefore many values arise in the context of the whole-life cycles of individuals with family responsibilities, rather than in the context of an individual over a short period. A large set of social values arises for many firms from the rearing of families in established home 'territories'.

Social values as well as instrumental arise also from improved relations with other workers. Some farmers gain social value by providing training. Many feel they gain social as well as intrinsic values from conservation of soil, water, woodland, etc. 'Leaving a farm more productive than when he found it' has long been valued highly by many farmers.

Table 12.1 lists what appear to be the important goals of farm firms and the types of value from which they arise. The classification of values is that used by Gasson (1973). Of course no general list of goals can detail well the goals of individual firms in particular communities and cultures, nor show the dynamic changes during life cycles and the development of firms. The list in Table 12.1 does, however, indicate goals and values that may explain many farmers' actual decisions better than do simple economic measures. (In other words, to give some examples: Why they continue to farm despite comparatively low money incomes and high variations in incomes from year to year. Why, at high costs, farmers continue to buy land for their sons. Why some farmers accept high risks and uncertainties to secure higher incomes, but others accept much lower incomes in order to avoid high income fluctuations. Why some farm families will emigrate from home 'territories' and some will not. Why some will continue to follow traditional systems of production while others change. Why some will excel in particular biological or work achievements.)

A general list such as that in Table 12.1 cannot show the relation between progress towards one goal and progress towards, or regressions from, another. (For example, higher average net incomes may be inconsistent with lower variations in net income flows, easier physical work, and happy family relationships.) Such relationships depend on particular value patterns, circumstances, plans, and management.

12.1.2 Assessments

Goals are particularly difficult to measure. Many are not stated in any quantitative terms, nor explicitly related to time periods. (For example: How low should the year to year variations in incomes be? What is the value of continuing in the home 'territory' now and in the future?) Some goals and values, although important to some individuals, may seldom or never be stated by them in verbal

165

Table 12.1   Goals of farm firms, and related values.

Goals	Main types of value to individual firms[a]
1. High average annual net income flows	
Returns withdrawn	I, Sf
Additions to net worth	I, Sf
2. Time for non-agricultural activities	
Earning	I, Sf
Other	S, A
3. High direct satisfactions from	
Leisure pursuits (e.g. fishing)	It, A
Rural living, including living in home area as against migration	It, A
4. Low variations in net income flows	
Seasonal	I, Sf
Year to year	I, Sf
5. Low liabilities in relation to net worth	I, Sf
6. Easier physical work, and better conditions therefor	It, S
7. Conservation of soil, water, wildlife, landscape	I, S, It
8. Rearing a family carefully and well in rural area	Sf, E
9. Training of future workers, including managers	I, S
10. Happy family relationships	Sf, I
11. Good relationships with hired workers	S, I
12. 'Belonging' in the community	S
13. High social status	S
14. Distinction as a breeder, feeder, grower, or for 'high farming' generally	S, E
15. Distinction as a 'traditional' farmer	It, S
16. Self-respect for doing a worthwhile job	E
17. Self-respect for being creative	E, A
18. Self-respect for using special abilities	E
19. Personal growth and achievement	E
20. Status as a landowner	S, E
21. Holding a large area of land	Sf, I
22. Enjoyment of rural tasks	I, A
23. Independence; freedom from supervision	I
24. Personal control	I
25. Good adjustments to life cycle changes including	
Transfer of assets and responsibilities to young	I, Sf
Retirement of old with adequate and secure income	I, Sf

Source: Based partly on Gasson (1973).
[a]Coded as follows: A, aesthetic; E, expressive; I, instrumental; It, intrinsic; S, social; Sf, social (family).

terms (e.g. conservation of wildlife, 'belonging' to a community, personal growth and achievement, freedom from supervision).

Therefore many assessments of goals can be based only on studies of (a) the actual decisions made by firms as shown by their results, as compared with the alternative results that might have been achieved; (b) the preferences expressed by firms for particular alternative plans, within a range of plans that have been based on various assumptions about the relative importances of different goals. Thus if during a particular period in a particular locality, taking high risks leads to higher average annual incomes, then farmers' actual decisions, or expressed forward preferences, help in the assessment of the values they place on higher average income levels as against avoiding income variations. Plans with careful crop rotations, natural biotic controls, and soil and water conservation may similarly have their probable outcomes compared with other plans maximizing higher apparent profits in the short run, so that farmers' values may be assessed.

In practice such assessments of values may not be satisfactory for several reasons:

(a) Actual past results of individual firms cannot be validly compared with results of plans drawn up with hindsight for some average local firm.

(b) Choices for the future may not be reliable, because decisions actually carried out may differ widely from preferences expressed at one particular time.

(c) By how much over a run of years 'profits' are affected by some values (e.g. of conservative rotations) cannot be reliably estimated. Other values (e.g. of being creative, of enjoying rural tasks) may raise or lower 'profits' by various amounts, the wide range being determined by how production decisions are affected.

(d) Assessments by comparisons are based on the assumption that all production decisions by individual firms are completely rational. For some individuals, this may not be a valid assumption.

(e) Measures of 'profit' are themselves liable to mislead.

12.1.3 Measures of 'profit'

As already noted, farm firms usually provide labour, management, use of net worth, risk and uncertainty bearing, and often land use. The composition of this whole can be altered to some extent by harder work for longer hours in some seasons, by more careful management, by saving more, by bearing more uncertainties. But to 'cost' the individual items separately, or to impute separate shares of net farm income to them, would require questionable assumptions. A common practice is to count all family labour, other than that of the farmer and his wife, as hired labour cost and then to deduct from *net farm income* the estimated value of the capital use (interest on asset values) and call the remainder the farmer's (and his wife's) *labour and management income* (or *earnings* if the value of the produce used by the family, and of the annual use of the farmhouse, has been included in income). Such income (or earnings) includes returns for risk and uncertainty bearing. Another practice is to deduct from net farm income the 'estimated value' of the farmer's (and his wife's) manual labour and call the remainder *interest on managed capital*. This also includes returns for risk and uncertainty bearing. No accounting methods can arrive at reliable measures of the 'pure profit' of economic theory. Therefore, when alternative farming plans include differences in farm firms' own inputs (e.g. more management and uncertainty bearing), the resulting general measures of 'profits' must be carefully interpreted. Otherwise they may mislead.

All measures of 'profit' require valuations and these cannot be precisely reliable (Section 10.5.10).

Within localities, comparisons between measures of the 'profits' of individual firms are commonly not valid unless the definitions of these measures are standardized to suit the purposes in mind. But this requires valuations of (i) farms and their rentals; (ii) manual labour not paid cash wages; (iii) products consumed; (iv) uses by households of farmhouses and gardens, vehicles, and stores. The costs of all borrowing are sometimes excluded.

12.1.4 Differences in 'profits' of example groups

Comparisons of 'profit' measures for our example groups are further complicated (e.g. by differences in money currencies, land tenure, and farm finance). The comparisons in Table 12.2 are of *net value added* by farm firm

Table 12.2   Net value added and its distribution: average farm, example groups, 1978–9.

Groups	Net value added per Hectare	Net value added per Man-equiva-lent	Rent of land	Return on other assets @10 per cent	Labour Hired	Labour Farmer and wife	Remain-der[a]
	US$	*US$'00*	*US$ per US$100 net value added*				
Aberdeen	376	193	17	26	27	16	14
Norfolk	587	220	22	20	35	4	19
Cortland, NY	378	212	17	26	24	11	22
Illinois	396	567	54	16	2	16	12
Saskatchewan	78	255	19	24	3	21	33
Louisiana	197	210	37	20	15	11	17
Iloilo	335	5	19	6	32	15	28
Aurepalle	107	3	20	10	12	5	53
Western Malaysia	363	7	5	33[b]	0	49	13

[a]For management, risk and uncertainty bearing, and taxation.
[b]Including all return on rubber trees.

operations (on an owner–occupier basis). This measure includes returns for the use of land and other assets and for all manual labour, as well as for management and risk and uncertainty bearing. Income and property tax returns for governments are also included, and cannot be well enumerated separately.

Net value added varied widely both per hectare and per man-equivalent (Table 12.2). The distribution to the various contributing inputs can be only approximately assessed but certainly differed greatly between the groups because (i) the biological and work contributions of particular inputs differed; (ii) the supplies of them in the respective socioeconomic systems differed; (iii) the values and goals of farmers differed. (For example, in Illinois the contribution of the fertile land was relatively great and land-use rights were in comparatively short supply. In Saskatchewan, other assets and risk and uncertainty bearing were especially important. In Iloilo hired labour for transplanting and harvesting rice had a larger share; and use of assets other than land had a much smaller share. In western Malaysia large shares were for the investment in rubber trees, and the smallholder's continuing work in tapping them and treating the latex.)

12.2 UNCERTAINTIES AND RISKS

The decisions of an individual firm should ideally be based on a body of knowledge much of which is *specific* to the sites, equipment, management abilities and energies, values, and other attributes of the firm. Much of the knowledge should be about *future* weather, values, prices, and other variables (Table 6.1 and 9.4). In practice this ideal body of knowledge can never be complete. Our consideration of the biological, work, and farm-economic sub-systems has shown that all may differ in many ways, and not least the abilities and energies and values of individual managers. Moreover, socioeconomic systems differ between countries and areas and change over time in ways that affect farm-economic sub-systems (e.g. prices, taxes, and subsidies change).

What is not evident from Tables 6.1 and 9.4 is that many variations in output–input relationships are smaller for firms with able, prompt management. Prompt use in unfavourable weather conditions of exceptionally short periods for early tillage and sowing, skilful control of weeds, pests, and diseases in accordance with rapidly changing conditions, wise use of irrigation, prompt harvesting, good stockmanship – all these reduce variations and make outcomes more certain. Good management of sales when demand is generally low may help to keep the total market value of production sold higher than it otherwise would be.

Some future possibilities of unfavourable events can be converted into known costs by paying insurance charges. Damage from fire, hail, and road accidents are examples, because the insurance companies have assessments of the risks that are well based statistically on past experiences in particular areas or for particular groups of people. Risks of low crop yields, because of drought in a particular area, may also be statistically assessed as a basis for government crop insurance schemes. But almost all decisions that farm firms make have to be based on subjective judgements. These may be helped by experimental data, past experience, studies of past price changes and their determinants, but seldom are the directly applicable statistical probabilities available.

12.3 RATIONALITY – LOGICAL PROCEDURES

Uncertainties and lack of definition of values are serious obstacles in rational decision making. Formal and informal education help to overcome these obstacles if they induce farmers to use such knowledge as will narrow their uncertainties, and to think more clearly about their own values and goals (Section 20.2.2). But inevitably, many important values are not well expressed nor measured directly, and some uncertainties remain. To narrow them further might cost more time and money than any resulting improvement in the 'quality' of decisions would be worth. The procedures in rational decision making therefore allow for uncertainties, and help to clarify and define values and goals.

Another feature of rational methods is that they pay attention to all *relevant* components of the biological, work, farm-economic, and socioeconomic systems, and all relevant values and goals of the decision makers themselves, whether well defined or not. Many of the interrelationships within and between systems are important, so that procedures that ignore all but a few can be seriously misleading even though, in the end, only a few may have 'key' importance for particular decisions. And the various values are commonly closely interwoven. Some choices can of course be well made even when only a few goals and system components are taken directly into account, but what are often called the 'backgrounds' to them must be well enough understood. Careful judgements must be made of (i) what should be brought into direct account, and (ii) what can safely be considered as 'background'.

Judgement must also be made of the length of time over which the components and values should be considered as affecting particular decisions (Sections 11.2.2 and 11.3.1).

The methods that have been developed to help farm firms in rational decision making aim first to provide measures of instrumental values (such as 'profit' and variations in 'profit'). Social, expressive, intrinsic, and aesthetic values are used in some methods to narrow the range of alternatives to be compared, and so make the comparison of instrumental values more realistic. (Some possible choices of product would be excluded because, although they might provide higher 'profit', their social, expressive, and intrinsic values for a particular farmer would be low or negative – e.g. milk production to a middle-aged farmer who has never before had the twice daily tasks of dairying. But all methods rely on assessment of the alternatives by farm firms themselves.)

Interpretations of firms' actual decisions may be inadequate for two reasons. The first is obvious. Only a few simple measures such as those of 'profit', or labour productivity, or crop yields may be used, so that several important values and goals are not well considered. The second is argument in a circle. For example, if low profits are explained by low labour productivity, and this is explained by the values placed on (i) leisure pursuits and (ii) avoiding the extra liabilities needed to secure adequate equipment, then no room remains for any further explanations. But in practice, at any one time, many farmers are themselves making new decisions about (i) and (ii) and other matters, because their previous decisions were not satisfactory.

Many farmers in all countries can be helped not only to narrow their uncertainties and define their goals better, but also to think through to better decisions. Interpretations should therefore be based on how far methods could be made more rational, as well as on how far uncertainties have been narrowed and values defined.

12.4 METHODS TO HELP RATIONAL DECISION MAKING

12.4.1 Basic economic principles

The scope of this book cannot include a detailed presentation of the basic 'economics of the firm'. But the methods that are introduced here are related to this basic logic (Table 12.3), so a brief summary of it may be useful.

Combination of inputs (choice of proportions in the use of factors) The simplest kind of choice is about how much of one factor (productive service) to use if

Table 12.3 Economic principles basic to methods that help rational decision making by farm firms[a].

	Principles related to					
	Marginal rates of			Marginal rates of substi-	Time pre-	Aversion to risk
Methods	Trans-formation of inputs	Substi-tution of inputs	Size of business	tution between products	ferences and dis-counting	and uncer-tainties
Partial budgeting	D	D		D		I
Investment appraisal						
Annual use costs	D	D			D	D
Valuation of net revenue flows	D	D			D	D
Whole farm						
Comparisons	I	I	I	I	I	I
Planning	D	D	D	D	E	E
Size of business						
studies	D	D	D	D	D	E

[a]Coded as follows: D, principles help to determine the most useful data to use; I, principles are especially important in the interpretation of results; E, both D and I apply, according to particular methods used.

171

inputs from all other factors are kept constant. The output–input relationship is fundamental and logic as well as observation leads to the conclusion that, 'If the quantity of one factor is increased by equal increments, the quantities of other factors remaining fixed, then the resulting increments of product will decrease after a certain point.' This is sometimes called the 'law of variable proportions', and sometimes the 'law of diminishing returns'. Figure 11.3 shows an example of its operation. The total product curves relating output of potatoes to inputs of nitrogen were not linear but curved, because the marginal product of the increments of nitrogen decreased throughout, and indeed became negative. They were additions to the nitrogen already available in the soil.

Suppose prices of nitrogen and of potatoes are such that 1 kg of nitrogen has the same value as 7 kg of potatoes. Then more nitrogen should be applied if the marginal product is more than 7 kg of potatoes per kilogram of nitrogen; and less nitrogen, if the marginal product is less than 7 kg (Figure 12.1). At the equilibrium point the marginal product has the same value as the marginal input.

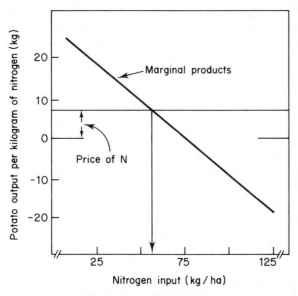

Figure 12.1 Equilibrium point in use of inputs of one factor, as determined by output–input relation (marginal products) and price of factor in terms of product. Output–input relation from Figure 11.3 (K_2O at 112 kg/ha); price ratio from market data.

The basic principles are as follows:

(a) The marginal rate of transformation between any factor and the product (the marginal product of the input $-Q_o/Q_i$) should equal the price ratio between the factor and the product (P_i/P_o).

(b) The marginal rate of transformation should be diminishing at the equilibrium position. This principle emphasizes the importance of knowing (i) the level and slope of the output–input relationship; (ii) the values per unit of inputs and outputs; (iii) the variabilities of (i) and (ii).

Almost always the variations in more than one input have to be considered because complementary relationships are common. Thus, in relation to Figure 11.3, we should know the marginal products of K_2O as well as those of nitrogen when we try to choose a 'balanced' fertilizer, and when we try to adjust to the changing prices of potatoes and costs of nitrogen and K_2O.

Figure 12.2 represents diagramatically the 'surface' of the output–inputs relationship. On each contour on this surface, there is logically one point where the combined cost of nitrogen fertilizer and K_2O is least. The basic principle is that to secure least cost production of the output represented by an iso-product contour:

(a) The marginal rate of substitution between two factors should equal their price ratio. (The 'prices' are cften alternative use values.)

(b) The marginal rate of substitution of one factor for another should be diminishing. (Thus along the iso-product contour, Q_{o_2}, each succeeding kilogram of extra nitrogen is a substitute for less and less K_2O, and each succeeding unit of extra K_2O is a substitute for less and less nitrogen.

If the price per kilogram of nitrogen is, say, twice the price per kilogram of K_2O, then the least cost is where 2 kg of K_2O substitute for 1 kg of nitrogen and vice versa. Such a substitution rate is indicated by the slope of the iso-cost lines (Q_{i_1}, Q_{i_2}, Q_{i_3}, ...) in Figure 12.2.

The least cost points on the iso-product contours can be joined by a line that is called the least cost 'scale line' or 'expansion path'. The same basic logic that applies to the one variable factor applies along the scale line to determine the equilibrium point where the difference between the total value of output and the total value of the two variable factors is maximized.

The same basic principles apply when more than two factors have to be considered.

Size of business This should logically be determined by (i) proceeding upwards along a least cost 'expansion path', inputs from all factors (except perhaps the firm's management, which is limited) being increased in accordance with the above basic principles; (ii) stopping when the equilibrium point is reached.

Choice of products In this too, marginal principles provide the basic guidance:

(a) The marginal rate of substitution between two products (quantities for quantities) should equal their price ratio.

(b) The marginal rate of substitution of one product for another should be increasing. (Thus in Figure 12.3 M'M" for each succeeding extra ton of beef the opportunity cost is an increasing amount of barley.)

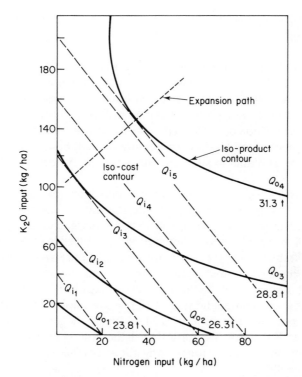

Figure 12.2   Expansion path for use of two variable inputs as determined by relation of output to these inputs, and by cost contours, Aberdeen area. Basic output–input relations as in Figure 11.3; iso-cost contours based on prices of nitrogen and potassium fertilizers in 1981.

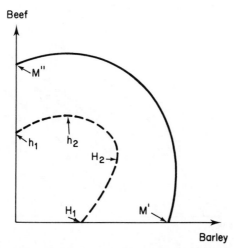

Figure 12.3   Diagram of output–input relations (production possibility curves). Short-term in modern agriculture, M' M'; long-term pre-modern, starting at H_1 or h_1, with complementary relationships to H_2 or h_2.

174

(c)   Each enterprise should contribute a positive surplus to the firm's total surplus.

In some circumstances the relationships between products are complementary. Thus in Figure 12.3, if farm plans have been ill developed, both more beef and more barley could be produced (e.g. by improving rotations, and seasonal use of labour). Historically, integration of animal with crop production made possible a long term production possibility curve with progress from H_1 to H_2, or h_1 to h_2, and achievement of the H_2–h_2 curve, before modern biological and equipment inputs became available to swell out the curve still further.

Time preferences and discounting   Investments alter the flows of inputs and outputs to be expected year by year into the future. Therefore to be able to compare alternative investments, and to apply marginal principles, future net revenues year by year have to be summarized and valued in some way that makes them comparable. A logical way is to discount future values back to present values by using a compound interest rate. This rate is a measure of the cost or disutility of waiting – of giving up current satisfactions or uses for future satisfactions (e.g. a compound rate of 10 per cent implies that £259 in 10 years time has a present value of £100). Many value judgements determine such discount rates. But here we may note that all the basic logic of marginal principles set out above applies in decision making about dynamic changes, provided that

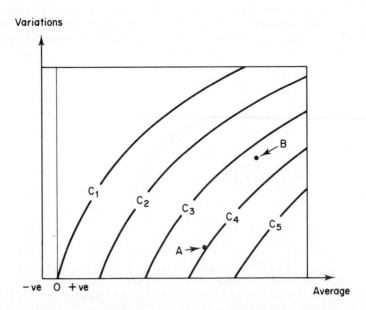

Figure 12.4   Diagram of indifference curves (C_1 to C_5) relating satisfaction to average revenues and year to year variations in revenues. If variations are expected to be high, higher average net revenues are required in compensation.

(i) all prices or values are discounted back to present values; (ii) the present total values of the streams of future net revenue are all positive.

Aversion to risks and uncertainties   The simplest conception of this is in patterns of indifference curves. Thus in Figure 12.4, for an individual decision maker, the curves suggest combinations of (i) average net revenue and (ii) year to year variability in net revenue (degrees of subjectively judged risk) that he or she would value as equally satisfying or dissatisfying. A production and investment plan that had the combination represented by A would therefore be preferred to that represented by B, even although the average net revenue would be forecast as higher with B than with A.

Because of aversions to uncertainties total assets are often limited. So progress up 'scale lines' is constrained and choices of product are affected. A basic principle remains, however – that the forms in which and the purposes for which assets are used should be such that the *marginal* returns on assets are equal from all purposes.

Relation of methods to basic principles   The methods introduced in the following pages of this chapter are listed in Table 12.3, with indications of the principles that should underlie each.

12.4.2   Partial budgeting

This method of helping farm firm decision making is the simplest and most used. The alternatives considered are changes from the existing choices of products and combinations of factors. The changes should not (a) require increases in equipment, land, regular workers, or other factors that should be regarded as 'fixed' because, if they are altered, changes in whole farm plans rather than partial budgets should be considered; (b) require increases in variable factors that will not be available or cannot be financed; (c) result in greater or smaller production of feedingstuffs as intermediate products without assessments being made of the consequences.

The changes are therefore 'marginal'. The quantities are essentially derived from judgement of output–input, input–input, and output–output relationships. The choice criteria are judged from prospects for market prices or other values to the firm (e.g. of produce used in the farmhouse).

An example is set out in Table 12.4, and this illustrates not only the method of partial budgeting but also some of the important points from 12.1, 12.2 and 12.3 that apply to all methods.

This example shows the partial budgeting for changes in our Aberdeen area reducing the spring-sown barley (cultivar Midas) on a farm from 75 ha by 10 ha. This 10 ha would be sown in autumn, to a new winter barley (cultivar Athene). The instrumental value measured is the £361 per hectare which is the difference expected between (a) the money value of the Athene grain and straw yields, *plus* reductions in costs, and (b) increases in costs *plus* the money value of reductions

Table 12.4   Partial budget for substitution of autumn-sown for spring-sown barley: 10 ha 'Athene' for 10 ha 'Midas', Aberdeen area, August 1981.

	£/10 ha
Additional benefits and reduced costs	
Grain from Athene, 47.0 tonnes @ £95/tonne	4465
Straw from Athene, 26.0 tonnes @ £25/tonne	650
Saving on Midas seed, 1.9 tonnes @ £105/tonne	200
	5315
Increased costs and reduced benefits	
Additional 20 kg nitrogen/ha and its application in autumn	102
Athene seed, purchased 1.9 tonnes @ £200/tonne	380
Additional fungicides and their application	172
Loss of grain from Midas, 40 tonnes @ £95/tonne	3800
Loss of straw from Midas, 20 tonnes @ £25/tonne	500
	4954
Net gain	361

in Midas yields. Such a single figure measure is easy to calculate. Indeed, many calculations of this type are done on the backs of envelopes, or by mental arithmetic.

But we should note the many judgements made, the uncertainties about them, and how far these might be further narrowed:

(a)   Because Athene is a new cultivar the experimental evidence of its grain and straw yield as compared with Midas yields is very limited. Judgements must be made in using what few data there are for the budget for the 10 ha site in a normal year.

(b)   Prices of grain are uncertain, including prices for possible sales for seed and of Midas for malting.

(c)   Values of straw are subject to big errors. Differences in the barley roots and stubble to be ploughed in are ignored.

(d)   The fungicide costs are uncertain because disease incidences cannot be well predicted.

(e)   Some drying costs may be avoided.

(f)   Other types of uncertainty arise from questions about skills and timeliness in management, particularly in relation to (i) the autumn soil preparation and sowing for Athene, (ii) nitrogen fertilization and harvesting of Athene, which seems more liable to lodge, shed seed, and suffer bird damage.

(g)   The sowing of *spring* barley *may* be more timely because the 10 ha are sown to Athene in autumn.

(h) There are also uncertainties about the actual alternative use value of the extra assets required during the October 1981–September 1982 period.

Thus, partial budgeting can serve to make clear the types of judgement that are being made, and to raise certain questions about goals and values, and about rationality. In this example, the farm firm might well at the end consider: what values to place on the additional net income and its variability; what are the 'social' values of easing spring work but increasing July/August work; what increase in 'social' and 'expressive' values would result from higher barley yields and higher soil fertility, and what decreases from greater fungal disease risks. And one or two deficiencies in logic might be apparent. (For example, if more grain and straw is to be harvested per hectare, should there not be allowance for extra fuel and other variable machine-use costs in harvesting and drying?)

This example indicates also several reasons why time is required for the narrowing of uncertainties at experiment stations and on farms, and why some farmers are earlier than others in making changes. (See also Section 20.2.2.)

12.4.3 Investment appraisal – 'lumpy' inputs

A few lumpy factors can, in some circumstances, be usefully considered in another type of partial budgeting (e.g. different types of tractor; small plantations or orchards; pig and poultry production units relying on bought-in feeding stuffs). Such factors will inevitably have some effects on the farm-economic system, as a whole, but these may be small enough to be treated as 'background'.

Comparison of annual use costs is the more common of the two basic methods for this type of partial budgeting. Thus two forage harvesters for grass to be ensiled in our Aberdeen area could be compared on the basis of their estimated annual costs (Section 8.2). These are of three kinds:

(a) The costs that, after the harvesters are bought, are unavoidable, and 'fixed' because they do not in total vary considerably with any changes in the area or tonnage of grass harvested. These have been called the DIRTI five – depreciation due to obsolescence and rust; insurance; rent or housing; taxes; and interest (capital use cost).

(b) Variable costs that vary with area or tonnage. These include other forms of depreciation and repair costs.

(c) Associated costs that also vary with area or tonnage. These include costs for labour and fuel and other tractor use costs.

But again, there are uncertainties. Depreciation and repair bills depend on management as affecting both rust and 'wear and tear'. The rate of obsolescence has to be guessed. The alternative use value of the additional capital required for the more expensive 'precision-chop' machine depends on other aspects of the farm economy and will vary in the future with prices. The greatest difficulties are in trying to measure the instrumental values in the faster rate of work and the higher quality of silage made possible by the 'precision-chop' machine. Estimates can be made of all the additional benefits, but they would vary according to weather, and assessments of quality differences and their effects. Closer estimation would lead us into whole farm planning.

Different farmers would also still put different valuations on certain other aspects of the two machines. Some would derive substantially greater intrinsic, social, and expressive values from the faster work, easier management, better products and 'up-to-dateness' of the 'precision-chop' machine.

Valuation of a flow of future net additional revenues is the other basic method (see Section 12.4.1 above). Thus the killing off of the old rubber trees in a small section of a smallholding or plantation in western Malaysia and their replacement by young trees of a new rubber cultivar could be compared with continuing the tapping of the old trees. The essence of the method is to make the best possible estimates of the initial investment costs and the inputs and outputs year by year into the future: then to compare with the estimated future inputs and outputs assuming there is no replanting. The method calculates either (a) the total *present value* of the net revenue flow by dividing each year's figure by the corresponding compound interest figure (this is the *discounted cash flow* procedure) or (b) *the internal rate of return* (this is the interest rate that would result in the total present value being calculated as zero). When alternatives differ widely in initial cost, the internal rate of return procedure is easier to interpret.

But again both procedures require subjective judgements because there are uncertainties about (i) quantities (technical coefficients) in the biological and work sub-systems and how they will change over time; (ii) future prices and price relationships; (iii) what discount rate to use. This discount rate can depend on the costs of borrowed capital (Section 18.4) or on the use values of the capital invested in alternatives that are not being compared. If the net revenues are single figure estimates, then the discount rate should perhaps be higher to allow for the uncertainties. The internal rate of return procedure does not entirely avoid this difficulty, because comparisons of the yields of the two alternatives would not be valid if all costs of borrowed capital, use values of capital in other alternatives, and uncertainties in various alternatives were not borne in mind.

Both procedures also leave open the final choice between alternatives according to social and other values not included in the instrumental value measurements (e.g. the social importance of continuing income to an old smallholder).

12.4.4 Whole farm comparisons

Various measures of a farm firm's economy can be compared with similar measures for other farms so as to provide indications of changes that the firm might rationally make. In addition to 'profit' measures, various other measures are used (e.g. of choice of product, combination of factors, size). Thus Table 12.5 sets out some comparisons of a farm in our Aberdeen example group to the most

Table 12.5   Some comparisons of one farm firm to 'most profitable' farm firms: Aberdeen example group, 1978–9.

	Averages for 22 'most profitable' firms	Individual firm	Shortfalls of individual firm[a]
			per cent
Size			
Land area, adjusted (ha)	112	102	9
Labour, manual (£)	7 736	5 660	27
Capital, tenant's, total (£)	80 773	42 811	47
Choice of product			
Cereals (per cent of gross output)	16	0	100
Cattle (per cent of gross output)	64	83	29[b]
Sheep (per cent of gross output)	7	5	30
Input mix			
Fertilizers and lime (£/ha)	44	20	55
Other variable costs (£/ha)	86	54	37
Animals, asset value (£/ha)	393	267	32
Gross output in relation to			
Land area (£/ha)	449	227	50
Work costs (£/£100)	337	226	33
Tenant's capital (£/£100)	62	54	13
Management and investment income			
In relation to tenant's capital (£/£100)	19	6	70

[a]Percentages by which the individual firm's figure was below the averages.
[b]Firm's figure exceeded the average.

'profitable' farms in the sample there. The diagnosis is that the particular farm firm had a relatively low grain area, low livestock numbers per hectare of grass, and low labour efficiency. If these were all raised, 'profit' (as measured by management and investment income) might be substantially increased.

Such indications may be strengthened if the relations of 'profit' to various other measures are studied by tabulation (as in Table 12.6), or by other correlation methods.

The diagnosis may therefore be useful to the firm. Understanding of the averages and variations of the measures made for the groups of farms is also valuable for those concerned with research, extension, and other parts of the socioeconomic system affecting agriculture (Chapters 13 and 20).

In practice, however, there are important limitations to the use of this method:

(1)   The farms in the comparison should (i) be sufficiently *similar* in many biological, work, and farm-economic respects; (ii) be sufficiently *numerous* for the correlation studies; (iii) provide data for *sufficiently long*.

Table 12.6   Relation of gross output and 'profit' to area of farms, inputs per hectare, and work costs: Aberdeen example group, 1977–8 to 1979–80 averages.

1	2	3	4	5	6	7
			Farms with above median			
	All farms	Area	Tenant's capital per hectare	Variable inputs per hectare	Area per £100 work costs[a]	Farms above all four medians
Number of farms	68	34	34	34	34	5
	hectares		*per cent of average in column 2*			
Area of farms[b]	99	156	100	95	135	137
	£/ha					
Tenant's capital	728	94	118	113	91	117
Variable inputs	135	93	117	128	93	126
	hectares					
Area per £1000 work costs[a]	5.98	109	93	97	120	109
	£/ha					
Gross output	414	97	110	113	93	124
Management and investment income	64	155	108	111	108	188
	£					
Management and investment income per £100 tenant's capital	8.7	166	92	99	120	161

[a]'Fixed' labour and machine use costs.
[b]Area of crops and grass plus rough grazing equivalent.

(2)   It is particularly difficult to secure useful comparisons for farms that already have relatively high 'profits'.

(3)   Many of the measures compared are *average* ratios (e.g. fertilizer costs per hectare). But the decisions to be made are essentially about *marginal* outputs and inputs.

(4)   There are the usual problems about valuations and measures of 'profit'.

(5)   These, and inevitable enumeration errors, can result in somewhat misleading tabular comparisons and correlations.

(6)   Measures of values other than instrumental values are not made directly.

(7)   The diagnosis of weaknesses can only be indicative. The individual firm

with its own unique set of resources and constraints has to study these and past results more fully, with the indications in mind. And it has to face its own problems and costs in making the transitions to greater strengths.

12.4.5 Whole farm planning methods

Purposes and essentials These methods are used when the changes would be too great for partial budgeting; also when a firm is considering a new farm. The detailed circumstances and values of individual firms are taken into account better than in whole farm comparisons. A variety of procedures can be used but at their best they all include the following:

(a) Listing of *'fixed' resources* that are *available* (e.g. land, regular labour, buildings). This list of 'fixed' factors shows some of the constraints on the plans.

(b) Definition of *'activities'* (e.g. growing of spring-sown barley, milk production) and measures of the *attributes* of these 'activities' (e.g. nitrogen fertilizer input), including the expected contribution of each activity less its variable costs). Also the probable variations in these contributions from year to year.

(c) Setting of *limits* that are judged desirable according to the firm's values and biological, work, and farm-economic goals (e.g. values in maintaining soil fertility and natural biotic controls and therefore the limits to changes in crop rotations or areas of particular intensive crops. These limits are other constraints on the plans.

(d) Combination of activities into a plan that would keep within the constraints, and maximize some measure of intrinsic value (e.g. 'profit', year to year reliability of 'profit', or some combination of these).

(e) Testing of this plan by assessing (i) its requirements of labour and equipment use seasonally throughout a normal year and probable variations in these in different years; (ii) the sources of money that would be available and its uses, including the cash flow during all periods when this might be seriously unfavourable; (iii) its 'sensitivity' to changes in yields, pests and diseases, and prices.

In practice, this sensitivity testing commonly includes a comparison with several other plans because the probable effects of some alterations at (a), (b), and (c) should be tested. Some of the limits at (c) might be reconsidered. 'Fixed' resources (a) might be altered (e.g. more land leased, better equipment bought). Some activities (b) might be redefined (e.g. after enquiry, agronomic practices might be improved; cows better nourished). Even some reconsideration of the measure of intrinsic value (d) might be recognized as desirable.

Which alterations are considered should depend on discriminating judgements because otherwise too many plans would result, and some would be misleading. Such judgements can be assisted by enquiry, research, and extension work (Chapter 20). Also judgements tend to be stimulated by the 'feedback' of results as planning progresses, and, in particular, reconsideration of 'fixed' factors may stimulate redefinition of some particular activities.

Definition of activities   This very important step requires more time perception and precision than any other in whole farm planning.

Definition should include all those attributes that may affect to any considerable extent the results of planning. Complete definition (e.g. including *all* detailed changes in timing with weather variations) is not practical. But uncertainties should be narrowed so far as seems meantime possible and worthwhile. Care should be taken to relate definitions to the biological and work systems of the particular firm, and its managerial abilities and energies.

Even with such care, definitions may be unreliable and inadequate. The reasons will be clear from a reconsideration of all the biological variables related to particular sites, agronomic practices, cultivars, animal genotypes, and animal nutrition, and all the variables in work systems (Tables 6.1 and 9.3).

Combination of activities   A variety of methods has been developed to help in this, ranging from intuitive planning to elaborate mathematical modelling and use of computers. This range cannot be adequately presented within the scope of this book, but the following notes serve to introduce it.

Intuitive planning is probably still by far the commonest method. Its logic is not a conscious step by step process, explicit and repeatable, but many of the farm plans that have resulted are confirmed as sound by other methods. This is because, at its best, it is based on intimate experience over many years of the biological, work, and farm-economic sub-systems of particular farms, and intimate knowledge of the firm's values and managerial and other constraints. Also it is a continuous process, each year's experience resulting in apparently desirable adjustments. At second best, it is based on a common 'follow-the-leader' process within localities.

But intuition may well not be adequate where the number of activities that should be considered is large, improved activities could be introduced, uncertainties are great, and values are ill defined.

Linear programming is a mathematical way of combining activities in plans that (i) do not break the constraints; (ii) do maximize the sum of net revenues from the activities.

The advantages are that (a) the logic is mathematical, proceeding step by step, explicit and repeatable; (b) the process can be carried through on electronic computers using standard programs *provided* that the activities are well defined, their attributes are well quantified, and the constraints are well chosen and defined; (c) because computers can do the long calculations very quickly, more attention can be paid to (b) and the results of sets of different assumptions can be obtained rapidly; (d) the mathematical framework (the basic matrix and its use) and the speed of computers can help to show up deficiencies in (b) (e.g. lack of sound data on labour requirements in busy periods) – some deficiencies may not of course be recognized; (e) in addition to the one maximizing plan for activity combination, the computer output can show marginal value products of 'lumpy' factors and other useful data to guide further planning.

The disadvantages are as follows:

(a) Because *linearity* is assumed, curvilinear output–input, input–input, and output–output relationships can be taken into the programming only by defining a sufficient number of activities (e.g. three or four silage production activities, each with different nitrogen-fertilizer inputs per hectare).

(b) *Divisibility* of actual activities into small units (e.g. individual cows) is possible, but computers cannot be instructed to accept no unit numbers within a range (e.g. between 0 and 50 cows). Therefore additional runs on the computer are necessary. Otherwise, plans would not be practical from a work standpoint.

(c) Possible changes in lumpy assets have, because of the *linearity* and *divisibility* assumptions, to be dealt with by an awkward range of definitions and re-definitions of constraints.

(d) Because the *independence* of activities is assumed, the actual biological interrelations between activities can be given effect only by defining 'composites' rather than separate activities (e.g. particular crops as one rotational sequence) or by requiring links between particular activities (e.g. that in the Norfolk area the spring-sown grain area should be at least equal to the sugar-beet area, because whatever this is it cannot be harvested early enough to be ready for any autumn grain sowing).

(e) Because the attributes of activities are stated as single value measures, simple linear programming may result in plans that are unacceptable because their risks are subjectively judged as too high. It is necessary therefore to impose various limits on risky activities. Certain goals in risk avoidance have to be defined.

(f) Only one intrinsic value is maximized – almost always the total gross margin in money terms. Other values have to be asserted as constraints or used at the end as criteria in subjective judgements of a range of plans.

(g) A plan formulated by linear programming is *static* and for the future. It would have to be put into effect by a transition from the present farm economy. But this transition is a dynamic process not guided in detail by the static plan. And this plan may even mislead if goals change, or assumptions should change, during the transition period.

Two other disadvantages of linear programming in many circumstances are as follows

(h)   Electronic computers are required and an organization and staff to ensure their proper use.

(i)   The mathematical process, data requirements, and reasons for results are difficult to explain to many farmers, and for this and other reasons, errors and omissions in activity definition and constraints may not be corrected early enough.

Programme planning is a process that avoids disadvantage (h) of linear programming and helps to avoid disadvantage (i). The whole of the arithmetic attempt to maximize the sum of net revenues is set out step by step by hand, starting with selection of the scarce resource that seems likely to be the most important constraint and the activity that would provide the highest net revenue per unit of that resource.

The disadvantages are that even though the calculations are time consuming

Table 12.7   Improvements on simple linear programming secured by other computer programming methods, and their disadvantages[a].

Method, and inadequacies of LP reduced or avoided	Remaining inadequacies, and special disadvantages
Monte Carlo, (a), (b), (c), (d), (e), (f)	(g). Not necessarily giving optimum plan. High computer use costs. Results may be difficult to interpret.
Integer programming (b), (c)	(a), (d), (e), (f), (g). Not necessarily giving optimum plan. Much time required.
Parametric programming (b), (c), (e)	(a), (d), (f), (g). Results are difficult to interpret if several resources and prices are varied.
Dynamic linear programming (f) (partly), (g)	(a), (b), (c), (d), (e), (f) (part). Discount rate has to be chosen. Price changes not included. Matrix tends to be large, very time consuming, and conjectural.
Dynamic programming (a), (b), (c), (d), (e), (g)	(f). Requires many matrices and models and more and better data, that are costly to secure.
Quadratic programming (e) (yields and prices)	(a), (b), (c), (d), (e) (inputs), (f), (g). Covariances of values of gross outputs should be known for various activity combinations.
Stochastic linear programming (e)	(a), (b), (c), (d), (f), (g). Many frequency distributions are required that are difficult or costly, or remain conjectural. High computer use costs.
Game theory strategy use (e), (f) (variations in income)	(a), (b), (c), (d), (f) (part), (g). Data requirements for range of 'years' usually cannot be met.

[a]For explanation of code letters (a)–(g), see text.

they may not result in maximizing net revenue, and (a)–(g) of the disadvantages of linear programming are not overcome.

Other computer programming methods do not have all these disadvantages or help partly to avoid them. Adequate descriptions can be given only in specialist publications. The more important and interesting methods that have been developed so far are listed in Table 12.7 with their special advantages and disadvantages. The choice of methods should depend on the particular problems to be solved and on the resources available. Mathematical *models that simulate* farm-economic sub-systems so as to provide comparisons of alternative dynamic changes are increasingly used for research purposes. Simulation models can be very useful in predicting cash flows and financial issues (Pack and Dalton, 1976).

12.4.6 Size of business studies

The important decisions that farm firms make about size of business can be assisted by all the whole farm planning methods applied to each firm's own unique resources and production possibilities, market conditions, and values and goals. They are particularly significant in determining the ratios of the inputs in the 'mix' and the amount of the 'mix' that is added to the managerial abilities and energies of each firm (Section 11.3.5).

In practice, size is measured in various ways – land area, cow numbers, labour

Table 12.8   Distribution of farms according to area.

	Norfolk[a]		Saskatchewan		Indonesia	Western Malaysia[b]
	1961	1979	1966	1976	1973	1973
	per cent of total farms					
Equal class intervals						
1st (smallest)	60	59	9	14	44	37
2nd	15	10	28	24	22	29
3rd	7	9	23	15	9	12
4th	5	4	14	21	8	8
5th	3	3	10	8	6	4
6th	2	2	5	7	3	2
7th	2	2	4	3	2	2
8th (largest)	1	1	2	3	1	1
Above 8th	5	5	5	5	5	5
	hectares					
Median	21	28	277	294	0.6	1.5
Range (of eight class intervals)	2–258	2–306	2.5–789	2.5–903	0.1–3.3	0–8.8

[a]Agricultural census holdings; some farms include more than one holding.
[b]By areas planted to rubber; smallholdings as sized in 1960, plus estates of more than 40 ha.

force, total capital, total gross output, and others. And how to define a farm firm's business has to be decided (Section 10.3). But always managerial inputs should be perceived, even although they cannot be well measured. These inputs become increasingly important as businesses become larger, and detailed biological, work, and farm-economic decisions increase in number and complexity.

Within localities and areas, the distributions of farm firms according to common measures of farm size is the result of decades and centuries of decision making. These distributions therefore provide some indication of the distribution of firms according to (i) managerial abilities and energies; (ii) values and goals.

Socioeconomic conditions are especially important in affecting decisions about size (Chapters 15, 16, 18, 19, and 20). Inheritance laws and practices; job opportunities that provide non-farm income; taxation; subsidization; land tenure and land reform laws; organization of marketing; education and extension; research and development – all these and other conditions determine the decision-making environment. And this along with the biological and work and human differences between firms determines which firms survive and which expand.

Table 12.8 presents some contrasts between four of our example areas, and changes over time in Norfolk and Saskatchewan.

12.5 BUDGETARY AND EXECUTIVE CONTROLS

Day by day management that is prompt and well informed and prompt 'feedback' of reliable detailed information for continuous decision making have an importance that can seldom be overemphasized. This is clear from variabilities in the biological and work systems, and in prices and other socioeconomic conditions. It is also clear from the uncertainties that complicate decision making itself and so make desirable the earliest possible assessments of actual results.

A wide variety of records is used to assist executive management and 'feedback' – bank accounts, other accounts, milk records, breeding dates, live-weight records, feed usages, and so on. And by expressing forward plans in budget form with important goals defined, management can be 'by objectives'. If the purposes of this are understood, another gain can be from the motivation of labour as well as of management.

12.6 MAN-MANAGEMENT AND ORGANIZATION

In practice, farm planning and executive management cannot be effective unless organization and labour management are suited to the biological and work sub-systems. They require prompt observation of many biological details; foresight; prompt and skilled responses; willingness to work long and hard at certain times; carefulness over controls of biotic factors; patience in the face of unfavourable weather and other conditions; other qualities. The requirements differ according to natural conditions and choices of production activities. (For example, milk production in the Cortland, NY, area requires a 'mix' of these qualities different from that required in much of the labour in the daily work of a rubber plantation, or in seasonal harvesting fruit or vegetables in areas specializing in these products.) Some work can well be done by large numbers of labourers, but the greater part of the labour in agriculture can function well only if it is in quite small teams that are intimately related to particular sites and herds or flocks and to the goals of individual firms (see also Sections 16.1.1·and 22.3.2).

REFERENCES

Gasson, R. (1973). Goals and values of farmers. *Journal of Agricultural Economics*, **24**, 521–538.
Pack, B. S., and Dalton, G. E. (1976). *Computerised Budgeting: An Approach*, Department of Agriculture, Reading University, Reading.

FURTHER READING

*Barnard, C. S., and Nix, J. S. (1979). *Farm Planning and Control*, 2nd edn, Cambridge University Press, Cambridge.
*Barlow, C., Jayasuriya, S., and Price, E. C. (1983). *Evaluating Technology for New Farming Systems: Case Studies from Philippine Rice Farming*, International Rice Research Institute, Los Banos, Philippines.
Cramer, G. L., and Jensen, C. W. (1979). *Agricultural Economics and Agribusiness: An Introduction*, John Wiley, New York.
Dalton, G. E. (1982). *Managing Agricultural Systems*, Applied Science Publishers, Barking, Essex.
†Dent, J. B., and Blackie, M. J. (1979). *Systems Simulation in Agriculture*, Applied Science Publishers, Barking, Essex.
Giles, A. K., and Stansfield, J. M. (1980). *The Farmer as Manager*, Allen and Unwin, Hemel Hempstead.
*Hardaker, J. B. (1979). *Farm Planning by Computer*, HMSO, London.
†Jensen, H. R. (1977). Farm management and production economics, 1946–70. In *Survey of Agricultural Economics Literature*, Vol. 1 (L. R. Martin, ed.), North Central Publishing (for the American Agricultural Economics Association), St Paul, Minn.
Ministry of Agriculture (1980). *An Introduction to Farm Business Management*, Reference Book 381, HMSO, London.
Norman, L., and Coote, R. B. (1971). *The Farm Business*, Longman, London.
*Rae, A. N. (1977). *Crop Management Economics*, Crosby Lockwood Staples, London.
*Renborg, U. (1970). Growth of the agricultural firm. *Australian Review of Marketing and Agricultural Economics*, **38**, 51–101.
*Roy, E. P., Corty, F. L., and Sullivan, G. D. (1981). *Economics: Application to Agriculture and Agribusiness*, 3rd edn, Interstate Printers and Publishers, Danville, Ill.

QUESTIONS AND EXERCISES

1. (a) If you were a farm firm, which of the goals in Table 12.1 would be most important to you, and which unimportant? (b) Which goals may become more important and which less important with changes in your age and family responsibilities?
2. Select a decision that could be made more rational by partial budgeting. Then (a) set out a partial budget; (b) give the sources of your data or estimates; (c) list the items about which you have substantial uncertainties.
3. Assume that the total output of agriculture in the area you know best is to be increased further. List under the following headings the changes that you would assess by whole farm planning: size of farms as measured by area; choice of products; 'lumpy' factors; regular labour; risk and uncertainty bearing; other management inputs; other inputs; loans to farm firms.
4. In the area you know best: (a) What records do farm firms keep to provide 'feedback' and so improve decision making? (b) What important items of 'feedback' are only recorded mentally? (c) What information that could be useful as 'feedback' do the least able and energetic farmers not secure?

Part V

The socioeconomic system and agriculture

Chapter 13

Introduction

13.1 SUMMARY OF FLOWS

Figure 13.1 summarizes very briefly the components of socioeconomic systems that are of most importance to agriculture. Farm firms and agricultural production are to the left of the figure, but markets for farm products and factors are to the right. In the centre are governments. Consumers, managers, and workers are on the right, but we recognize that those in farm households have especially direct influences on farm firms.

No attempt is made to show in detail the relations of consumers, managers, and workers to product and other markets, nor to governments. The lines

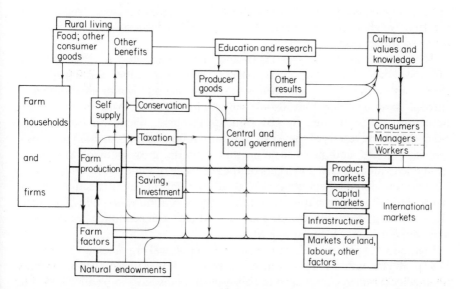

Figure 13.1 Socioeconomic system: simple diagram showing interrelations with agriculture.

connecting the 'boxes' indicate flows in both directions, unless arrows are shown.

The system as a whole can best be understood by starting in the lower left-hand corner and progressing upwards. The *farm factors* ('input mix') box receives flows from *natural endowments*. And natural endowments are affected by the decisions of firms responsible for *farm production* and *factors*. Natural endowments also provide *non-material benefits* in *country living* (see Section 12.1.1).

The *farm factors* 'box' also receives flows from *markets* for land, labour, and other factors and in turn makes demands on these markets according to the decisions of farm firms. These decisions are themselves influenced by the important flows into the farm factors 'box' of what may be regarded as 'producer goods' resulting from *research and education*, at the top centre of the figure (see Chapter 20).

Farm production is also influenced by demands for its outputs. In subsistence agriculture a major influence is from the farm household demands for *self-supplied* produce and also for non-material benefits. In commercial agriculture, prices of products as determined in *product markets* are important criteria. Indeed, the farm production–product markets flow can often usefully be regarded as the main axis of the system summarized in Figure 13.1.

But if demands are to be met, farm production must be able to use sufficient assets. It therefore depends on the 'boxes' labelled *capital (loan) markets* and *saving and investment* because these affect the opportunities of farm firms to borrow and invest (Chapters 10 and 11).

The 'box' labelled *infrastructure* is also important because it represents all those social arrangements that are required for effective operation of the system but are not included in the other boxes (e.g. for law and order, social and political security, weights and measures, plant and animal hygiene, communications, money and banking).

The opportunities of farm firms and the constraints on them are also influenced by *central and local governments*. Their controls and subsidies affect all the 'boxes' to the right of and below 'farm production'. Major effects are also through *taxation*. There may also be *conservation* controls. And very important influences can be exerted by governments through policies on *research and education*. Some results of these can be information and skills that are obviously '*producer goods*' adding to farm factors, and the great pool of knowledge available to managers and workers on the right side of Figure 13.1. *Other results* may be considered '*consumer goods*', providing immediate satisfactions within particular cultures, or in a more durable way to *cultural values and knowledge*, thus affecting decisions throughout the whole socioeconomic system (e.g. demands for higher living standards).

Figure 13.1 is useful in relation to all types of political organization because all have problems within and between all the 'boxes'. (For example, even in centrally planned economies, assets have to be supplied, loans have to be allocated, and prices determined.)

13.2 HISTORICAL PERSPECTIVE AND MODERN TRENDS

13.2.1 Before commercialization

In a few parts of the world there still exist self-sufficient communities made up very largely of farm households. They have cultural values and knowledge that support their biological, work, and farm-economic sub-systems. And their tribal, clan, and family education sustains these values and types of knowledge.

13.2.2 Population pressures and commercialization

In most parts of the world three developments aggravated the problems of such communities and led to changes in cultural values and socioeconomic relations. Land and other natural endowments became scarcer and more valuable when (a) human population pressures on them increased; (b) they could be used to secure additional income from sales of products because markets were widened; (c) production for sale could be increased by use of animal or mechanical power, and biological factors.

These three types of change started before or soon after civilization. (Sections 6.2 and 6.3). Once started, the development of agriculture depended increasingly on all those components of socioeconomic systems that are to the right (below and above) of the *farm production* 'box' in Figure 13.1, as well as on farm firms themselves. History is indeed much concerned with the elaboration of the 'boxes', with failures and successes in this, and with changes in them and fears of changes. And, as we noted, socioeconomic systems depend for their development closely on the development of agriculture. So as civilization progresses close interdependencies are increased and become more intricate. When civilizations decline, these interdependencies are disrupted.

Many modern controversies indicate the heavy importance of government policies as affecting agriculture, and the importance of agriculture for governments.

13.3 REASONS FOR DETAILED CONSIDERATION OF AGRICULTURAL PRODUCTION

Many economists tend to consider agricultural production only as part of a 'food industry'. They thus usefully emphasize the close interdependencies between farm firms and markets for both products and factors. They also emphasize the opportunities that some governments have to alter their policies on agriculture. Some biologists are keen to advance 'biotechnology' and suggest that agriculture will soon not be essential as a special form of organization for photosynthesis.

However, it is still useful for many purposes to consider farm firms and agricultural production in detail. Otherwise substantial mistakes may be made because of the biological, work, and farm-economic aspects of production on farms and the special interrelations with natural endowments, conservation, non-material benefits in rural areas, land markets, and security of future food supplies. Socioeconomic demands for photosynthesis on farms seem likely to increase for many decades.

13.4 PURPOSES OF LATER CHAPTERS IN PART V

The remainder of this part surveys briefly all the 'boxes' in Figure 13.1, starting with *product markets* (Chapter 14) and proceeding to *factor markets*. Some important parts of infrastructures (including government finance) are introduced in Chapter 19, and education and research in Chapter 20.

Social values and goals require consideration in Chapter 21.

QUESTIONS AND EXERCISES

1. List the markets and other institutions that can affect farmers' decisions.
2. Suppose a nation wishes agricultural production to be increased. List and briefly explain the kinds of action the central and local governments *might* take to help to give effect to this wish (e.g. reduce central and local taxes on farm products, factors, and incomes).
3. During the last 20 years, what have been the five most important influences of the socioeconomic system on the agriculture of the area you know best?

Chapter 14

Product markets and marketing

14.1 INTRODUCTION

Markets for products are essential to the development of agriculture. Without them (i) trading cannot develop between areas, regions, and nations, and so the choices of product by farm firms cannot be closely related to the biological conditions of particular sites; (ii) labour cannot specialize in non-farm activities so that what labour remains on farms can be more productive; (iii) agriculture cannot secure the purchasing power to buy fertilizers, machinery, and other goods and services required to raise yields and overall efficiency, nor can farm households buy consumer goods and services and so be stimulated to produce more farm products – furthermore, taxes cannot be paid that help to finance government services to agriculture; (iv) the biological variations from year to year in crop yields and livestock outputs cannot be so well evened out by storage and trading, and by saving money in 'good' years and dis-saving in 'bad'.

The interdependencies of agriculture and markets are shown by the range of foods sold retail in any modern big town. They are derived from many regions and nations and many crop species and cultivars, animal species and breeds. The basic farm products have had many services and some materials added to them (e.g. transport, processing, storage, packaging). In the USA a modern superstore commonly has some 7000 or more items of food well displayed for consumer choice. In many countries the total retail value of foods bought is double or more the value of foodstuffs sold off farms or imported.

The flows along the 'axis' of Figure 13.1 between *farm production* and *produce markets* and towards *consumers* and the flows between product markets and international markets are therefore very large, complex, and important.

14.2 DEFINITION OF 'MARKETS'

For our purposes the most useful definition of a market is 'a group of buyers and sellers with facilities for trading with one another'. How wide the group is considered to be depends on the functions or problems being studied (e.g from local auction markets to international markets).

Governments, and bodies such as marketing boards, can be buyers and sellers or can intervene in markets by various forms of regulation (e.g. import duties;

delivery quotas; licensing) or by subsidization and taxation. Some forms of intervention divide markets so that at one 'official' price level rationed produce is sold to consumers, while at another, there is exporting or importing, and at 'black market' prices uncontrolled supplies flow. Other trading or interventions express the demands of governments for stocks (e.g. for security of food supplies, or when they wish to raise the incomes of farm firms).

In some economies that are 'centrally planned', the interventions of government agencies may seem to make 'markets' very different from those in 'market economies'. But the basic functions that markets should perform are essentially the same in all 'socioeconomic systems'.

14.3   FUNCTIONS

The basic functions of product markets can best be understood if they are considered in four groups and in relation to the conception of the 'perfect market' (Sections 14.3.1–14.3.5).

14.3.1   Matching demands and supplies

In all countries, markets should

(a)   Assess and register through prices or in other ways (i) the demands and preferences of consumers of agricultural products, and variations and trends in these demands and preferences; (ii) short-term changes in agricultural outputs due to weather and other causes; (iii) longer term technical and economic opportunities and changes affecting agricultural production, processing, packaging, storage, transport, and distribution; (iv) political and social goals (e.g. security of national food supplies; nutrition of low income families; sufficient net incomes for farmers; price stability; inflation control)

(b)   Facilitate achievement of the economically most appropriate volume, type, location, seasonality, and quality of output of individual agricultural products, the methods of sale, and the marketing services attached to them

(c)   Govern the time, place, and manner of consumption of agricultural products and the use of processing and storage

If these funtions are not carried out well, the decisions of farm firms cannot be fully economic about choices of product (including times of production and qualities), intensities of production, storage on farms, and all other determinants of supplies, locations of supplies, and changes in supplies. Nor can the decisions of traders, processors, consumers, and governments determine well consumption, trading, and movements into and out of storage. The opportunities for production are not well taken, nor matchings of supplies to demands and demands to supplies well made. Shortages or short-term surpluses arise. Medium and long term shortages or surpluses may arise, indicating that agricultural resources are misused. Best use is not made of new technology on farms or in marketing.

Marketing arrangements in any particular area are tested, in part, by short term changes in supplies due to weather and biotic factors. If the market is wide and flexible enough, fluctuations in production will be largely off-set by increases (or decreases) in stocks in store, and by increases (or decreases) in 'exports' from the area, or decreases (or increases) in 'imports'. Consumption will also be adjusted, particularly consumption of perishable foodstuffs that are costly to store, or lose quality if stored. Seasonal changes in supply provide similar tests. Marketing arrangements are tested also by changing biological and technological opportunities (e.g. those in control of insect damage to fruits; refrigeration). Thus we can identify narrow markets that do not match supplies and demands well, and backward markets that delay innovations on and off farms.

Divisions and barriers within potentially very wide markets can be seen when information about national supplies is not fully and promptly communicated (e.g. in 1972 when shortfalls of the USSR's grain harvests were not sufficiently early understood in world grain export markets).

The main way of registering supplies and effective demands in markets and matching them is through prices. A brief introduction to the price mechanism is therefore given in Section 14.4. We should note here, however, three basic requirements:

(1) Supplies have to be promptly assessed in quantity and quality and considered in relation to effective demands. And prices have to be widely and promptly known if trading and storage are to help to determine prices well throughout a market. Much information has to flow quickly into and within the network of buyers and sellers. To provide this information organization and procedures are necessary.

(2) The desirable width and flexibility of markets also depend on organization, equipment, and procedures for transport, storage, processing, retailing, financing, legal and arbitration procedures, and other 'infrastructure' that helps to make transactions between buyers and sellers easier and more certain.

(3) Because the functions of markets are commonly held to include achievement of some political and social goals of governments, close studies are desirable of prices and their effects, and of any interventions affecting supplies, demands, and prices. Such studies are especially important where various short term goals may be inconsistent (e.g. lower retail food prices, with higher incomes for farmers; lower taxes; and high government subsidies).

14.3.2 Physical and economic efficiency in marketing

Markets should (a) achieve economic efficiency in the physical assembly, classing or grading, processing, packaging, transport, storage, and distribution of agricultural products, and in the finance and risk bearing related to these operations; (b) provide facilities for appropriate methods of sale and bargaining. This second big group of functions is essential to keeping down costs of marketing. It also supports the first group in helping to match supplies and demands. (For example, if products are economically assembled and classed or graded, assessments of the qualities and quantities of supplies are much easier and quicker.)

The functions in (a) are usefully considered separately as functions of the individual firms of buyers and sellers, and of those who finance them and provide transport and other factors to them. In essence, firms in marketing produce *space, time, and form utilities (benefits)*. These result from physical movements, storage, and processing which may be regarded as intermediate 'products' of firms. They make possible overall the effective use of supply opportunities, and the matching of supplies and demands. This is best understood by considering examples in the marketing of one 'commodity' such as 'wheat and flour' in Canada. The physical movements and related financial services are many and complex. The efficiency with which they are carried out depends on a whole 'structure' of firms ranging from small co-operative societies to large milling and shipping firms. Two major government agencies are the Canadian Grain Commission (responsible for grades and standards and statistics) and the Canadian Wheat Board (with overall responsibility for marketing the prairie wheat crop, including farmers' delivery quotas and prices for all exports). This is the structure that has evolved, over decades, as a result of various experiences, and changes in technological, economic, and political circumstances.

14.3.3 Sustaining and evolving for the future

This set of functions is notable because particular government policies and goals are liable to affect not only overall efficiency in the short run, but also longer run capabilities.

Research and development, training, and education are essential to raising economic efficiency in marketing as well as in agricultural production. Marketing firms of all kinds should, like farm firms, co-operate in facilitating these services and in using them. Some particular types of 'development' aim to raise effective demands for products by study of consumers' preferences and spending powers, by improvements in quality and attractiveness of products (e.g. by packaging and by advertising in various ways). Such developments are usually costly, but they can be economic if supplies and additional marketing services are well matched to consumers' longer run interests and effective demands.

Another reason for emphasis on the future is that in all countries some buyers and sellers are inclined to take narrow views of their own best interests. Their true long-run interests are essentially that (i) they can secure the 'input mixes' of management, information, transport, storage, processing, and other factors and finance that are necessary to produce economically the services for which there are (or soon will be) effective demands; (ii) components of the 'input mixes' are paid for or rewarded according to their alternative use values; (iii) probable future changes in effective demands for services, supplies and costs of inputs, technology, and government goals, are understood early enough and well enough so that difficulties and costs in changing 'structures' for marketing can be minimized. (See Section 20.1.2.)

14.3.4   Price and other controls for particular policy reasons

Increasingly priority has been given to the safeguarding of human, plant, and animal health. Other common examples of socioeconomic goals are (i) to keep retail prices down during periods of inflation; (ii) to secure equitable distribution during periods of scarcity (e.g. of sugar, fats, and meat rations); (iii) to make the returns to farm firms higher than they otherwise would be, in order to secure higher production from farms, or more equitable income distribution; (iv) to stabilize prices paid to farm firms; (v) to earn more foreign exchange from exports or save foreign exchange spent on imports; (vi) to promote lower cost marketing; (vii) to promote better quality produce and more services for consumers.

Social values underlie the definition of such objectives (see Section 21.8), but immediate political pressures from interested groups influence their timing and detailing.

The main methods that governments use in attempting to secure their objectives affect marketing functions through the following:

(a)   *The price mechanism*   (i) Its *free working* may be facilitated and made more effective by many different types of service to marketing firms (e.g. more, earlier, and better quality statistics; improved transport, telegraph, and postal services; improved credit facilities; reduced import or export duties; legal safeguards; better education and training). (ii) *Controls of prices* may be direct, mainly through legal maxima and/or minima, subsidies, duties or levies, fixing middlemen's margins and locational or quality differentials, and trading by government agencies themselves. The procurement arrangements of centrally planned economies are based mainly on delivery quotas at low prices into official marketing channels, and the fixing of margins for every stage along these channels and of legal rations and retail prices to consumers; but quantities produced in excess of the delivery quotas may be consumed on farms or marketed at higher and freer price levels. (iii) Controls may be *mainly through supplies*, by levies or quotas on imports, control of exports, restrictions on marketing, government trading and surplus disposal, and regulations introduced for health control reasons. (iv) Controls may be *mainly regulation of consumption* by rationing, which entails a procurement system to secure supplies and a distribution network to ensure that they are always available at the right times and places against ration demands. Advertising and extension programmes

may be used to influence consumption. Also *subsidization* (e.g. of school children, pregnant and nursing mothers, and low income and other nutritionally vulnerable groups).
(b)   *Taxation* by various methods.
(c)   Direct controls through laws and 'police powers' (e.g. hygiene and anti-pollution regulations; trade quotas; quality control of exports and imports; labelling regulations; other measures for consumer protection).

14.3.5   The 'perfect market'

This is a conception that helps to summarize the functions in marketing. It is an ideal that is virtually impossible to achieve in practice, because 'supplies' and 'demands' are so dynamic and sellers and buyers seldom can have enough precise knowledge. But the ideal is a standard against which realities may be judged. Although closely related to the theoretical conception of 'perfect competition' (Section 14.4.1), it can be used separately.

In a perfect market:

(a)   Prices do not differ between locations, times, or qualities and forms of product by more than the respective costs of transport, storage, and quality control or processing. In other words, the benefits of additional space, time, and form utilities would not be greater than their costs.
(b)   The efficiencies of the firms providing marketing services could not be improved by further changes (i) in their numbers and size and the overall 'market structure'; (ii) in their internal organization and management.
(c)   If the policies and goals of government are to depart temporarily from (a) or (b), the direct and indirect, short and longer term benefits and costs are correctly assessed.
(d)   Through prices and technological and market outlook services farm firms obtain reliable guidance of their decisions about production, storage, and processing on farms.
(e)   New economic opportunities in agricultural production are supported by changes in marketing, so that consumers have security of supply, health safeguards, wide ranges of choice, and other services for which there are effective demands.

14.4   THE PRICE MECHANISM AND PRICE CHANGES

Prices are essential in the economic arithmetic not only of farm firms but also of firms whose products are marketing services. And the consequences of price changes are far reaching. They include changes in choices of product and the use of resources in agriculture; changes in structures for marketing, marketing efficiency, international trade, and security of food supplies; and changes in the distribution of incomes; and the variabilities of incomes. An understanding of prices is therefore an important key to sound development of agriculture.

14.4.1 Basic theory

This is not complicated. Supply and demand are conceived as occurring during a short time period in a well defined market. They are 'equated' at the 'equilibrium' price. Any higher price would induce sellers to offer more for sale, but buyers to reduce their purchases. Any lower price would induce buyers to ask for more, but sellers to offer less. There are no restrictions on the numbers of buyers or sellers, nor on their information. Together with some other basic assumptions (see economics textbooks), these are the conditions of 'perfect competition'.

14.4.2 Supply schedules

To make use of basic theory, we have to consider how the decisions of farm firms will be altered by price, and so change the aggregate supply schedule. We have to return, therefore, to the values and goals of farm firms, their partial and whole farm budgeting, and their management. Table 14.1 lists the variables that are the main determinants of what farm firms will actually sell into a market. Some affect forward production plans and future supplies; others affect also the timing of sales (e.g. items 6, 11, 12, 16, 17, 18, and 19).

Table 14.1   Determinants of changes in the supply of a farm product (Y) in a market.

1. Prices, expected or assumed, of Y and of other farm products, including quality premia or discounts
2. Prices, expected or assumed, of variable factors used to produce Y and other products, and availabilities of these factors
3. Costs, expected or assumed, of 'lumpy' factors that could be used to produce Y
4. Effective rates of discounting over time
5. Subsidies on products and factors
6. Taxation rates and allowances
7. Output–input relationships over time, and input–input and output–output relationships
8. Changes in the technical and business knowledge and abilities of decision makers on farms
9. Risks and uncertainties attaching to all the above as affecting management needs, and the variability of net returns
10. Changes in the 'sentiment' of decision makers (Favourable and optimistic periods as against others)
11. Capital rationing – voluntary, by farmers
12. Capital rationing – voluntary, by lenders
13. Constraints by landowners
14. Constraints by governments or their agencies
15. Values and goals of farm firms other than 'profit'
16. Storage facilities and costs
17. Demands for stocks – judgements of gains or losses from holding stocks
18. Supplies and demands from other markets
19. Use value of Y on the farms where produced (e.g. of grain as feed; animals for breeding)

202

In addition to farm firms, other market firms (including government agencies) determine the supplies offered, particularly through items 12, 16, 17, and 18. Some important determinants of supply at any one time are also elements in demand (e.g. items 17 and 18). We have also to recognize that 'product Y' is made up of a range of qualities, and the proportions that each makes up of total supplies vary over time and locations (Section 11.2.5).

Thus in practice the current prices of a product are usually important in determining supplies now and later, but these have many other determinants. Because these tend to vary widely, supplies are not constant nor precisely predictable. All the complex details of supply determination are seldom captured in studies of markets and prices. The statistics and related information that would be required are too difficult and costly to secure.

But major influences can be measured and interpreted, bearing in mind the market's 'background' of other variables. Some studies of prices and markets attempt to measure the average precentage change in supplies for each 1 per cent change in product price. This is the *elasticity of supply*, indicating the 'slope' of the supply schedule. Many variables in the economies of various sizes and types of farm firm can be used to forecast their decisions, and these then aggregated. The results cannot be precise, because of all the uncertainties that we noted in Section 12.2 and other variables in Table 14.1. Other studies analyse past changes as a basis for forecasting future responses. But only some of the variables are measured; samples of time are short, and future variations (e.g. changes in technology) are not known with certainty. Moreover, some farm firms and some marketing firms (e.g. processors) take time to alter their decisions. So studies of the elasticity of supply attempt to answer questions that are wider than those that basic theory reasons out. The answers can be only indicative, and are different for the short, medium, and long term futures.

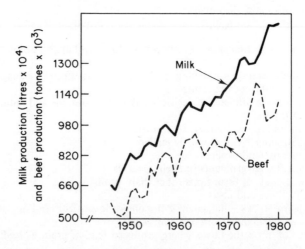

Figure 14.1   Production of milk and beef, United Kingdom.

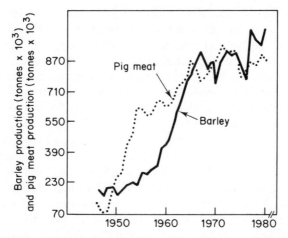

Figure 14.2   Production of barley and pig meat, United Kingdom.

The fullest possible understanding of all the determinants of supplies is, however, essential to sound marketing and price policy. For different products and markets, these determinants differ in detail so that the trends and variations in supply differ. Simple examples are contrasted Figures 14.1 and 14.2.

In 'centrally planned economies' production plans and product prices in the 'official' market are set for several years into the future. In 'market economies' the many determinants of production operate more continuously. Governments can cause discontinuities, but a basic aim is continuous adjustment.

One important type of such adjustment is to the timings of sales off farms of annual crops, and of movements into and out of store. Timings are determined in dynamic ways by prices, demands for consumption and for stocks, and information about supplies, demands, and prices. Inclusion in markets of buyers and sellers of contracts for future delivery ('futures') can assist such adjustments if there are many such traders and statistical information is satisfactory. Price uncertainties are then spread to those who would bear them, and all buyers and sellers have better guidance.

Adjustments of livestock production are dynamic because they are influenced so much by variations in feedingstuffs supplies, and conditioned by life cycles and gestation periods. Flexible pricing and adequate information are usually essential to satisfactory results.

14.4.3   Demand schedules

In basic price theory, the *price elasticity of demand* is conceived as the percentage by which purchases change when prices are decreased by 1.0 per cent. It measures the 'slope' of the demand schedule of a market. More satisfactions are secured from more consumption. But, per unit, these satisfactions diminish as more and more units of a product are consumed.

In practice, demand schedules are determined by many variables. Populations increase, and change in age composition. Incomes and income distributions change (Chapter 1). Many other influences (e.g. from religion, education, culture contacts, urbanization and industrial employments, advertising, packaging) also alter consumer satisfactions from purchases. Changes in supplies and prices of substitutes for particular products may be especially important (e.g. those of artificial fibres affecting demands for cotton and wool). Studies are made of these variables and other types of elasticity are assessed (e.g. (i) *income elasticity of demand* – the percentage difference in consumption with a 1.0 per cent increase in income; (ii) *cross elasticity* – the percentage change in the consumption of one product resulting from a 1.0 per cent increase in the price of another product). Actual future changes in consumer demands are, like those in supplies, difficult to predict reliably. Types of change that have important long term effects on demand are changes in technology as affecting storage, transport, and processing (e.g. the development of refrigeration, modern wine making). Such developments spread demands to areas previously with fewer buyers. They also tend to alter demands for different qualities.

Even more important in shorter run periods are changes in demands of buyers who are not ultimate consumers. The demands of storers, exporters, and governments can all be substantial. Some processors of foodstuffs produce non-food products (e.g. industrial alcohol). We have also noted that on farms there are demands for feedingstuffs that may be large when prices fall (e.g. for wheat as a feed).

14.4.4 Pricing when competition is not perfect

In some market conditions, *individual* sellers are faced with a downward-sloping price curve, so that the more they sell the lower the price (average return per unit), and the lower still is the marginal return per unit. In such conditions it seems rational for the *individual* seller to stop increasing supply at the point where marginal cost equals marginal return. Supplies may therefore be restricted to keep prices above what they otherwise would be.

Similarly, in some market conditions, *individual* buyers do not face the same price, however much they buy. Their marginal purchase cost rises with increasing purchases, so that the average purchase cost per unit also rises. It seems rational, therefore to restrict purchases so that the marginal value to buyers (perhaps for onward sale or processing) and the marginal purchase cost are equal. This can result in a high difference (margin) between the average value to them and the average purchase cost.

Thus the conditions that make the demand curves that *individual* sellers face, and the supply curve that *individual* buyers face, less than completely elastic (level) can be important. Important types of condition are considered in the following paragraphs.

Reduced numbers of sellers   The number of sellers may be reduced to few (oligopoly) and even to only one (monopoly). The number of farm firms is large in almost all markets. A few co-operative societies and other groups have achieved sole buying rights within their own areas and sell on their behalf. But they usually sell into markets with many other sellers with like products, or close substitutes. Oligopoly does exist amongst government agencies for some commodities exported into world trade (e.g. wheat, some dairy products), but such agencies do not have close control of production.

Monopoly is established for a few years through registration of 'breeders' rights' in some countries when a new crop cultivar is released, but competition from close substitutes usually keeps prices from rising very high.

Reduced numbers of buyers The number of buyers is sometimes few (oligopsony) and even only one (monopsony). In processing some farm products (e.g. sugarcane, sugar beet, tea) economies of scale are such that, in any particular locality, only one plant can operate efficiently. In some market structures, vertical integration (Section 11.7) makes supplies not contracted for difficult to sell (e.g. eggs and broilers). In some local auctions, buyers may be so few that they share supplies and so secure them at lower prices. In some remote areas (e.g. in eastern Malaysia) the number of buyers is few and they retail consumer goods and lend money as well as buy produce, at least partly because specialists would have insufficient total business. The most common single buyers are government agencies (including marketing boards, who received local monopsony–monopoly powers from government). Markets can be split by monopsonies that sell onwards at a higher price level for national consumption but at various lower price levels internationally.

International trade can be restricted or stopped by a variety of other barriers (Section 14.3.4).

Differentiation Markets can be split up into different parts when sellers differentiate their products by intrinsic quality, or branding, packaging and advertising. The demand curves for each seller may be virtually the same and level if buyers are not impressed, and therefore readily substitute for purchases from one seller those from another. But the curves may be raised and sloped if buyers acquire some special preferences for particular sellers' products.

Differentiation tends to be greater at retail stages and if different types and amounts of services are added to products. The practices of retailers, and related processors, packers, and storers, thus affect the demand curves for consumer goods and, indirectly, the demand curves faced by farm firms. The effects tend to be greatest where there is much innovation and vertical integration.

Other important causes of imperfect competition and unsatisfactory working of the price mechanism are inadequacies in the information available and in the understanding and use of information. Buyers can then lower prices to sellers. This seems often to occur in poorer countries where buyers are better educated and trained than farmers (Barlow, 1978). In all countries, 'margins' which are the gross returns (prices) for the services of particular types of middlemen are more difficult to observe and understand than are farmers' prices and retail prices. In other circumstances some sellers can raise prices to buyers (e.g. by misinformation about qualities or quantities).

Price mechanisms cannot of course work well if not enough investment has been made in the 'lumpy' inputs and infrastructure required to secure all the time, space, and form utilities for which there would be effective demands. (For example, if the transport system is inadequate, supplies will not be well spread from areas of low cost production. If storage facilities are inadequate, seasonal supplies and occasional years' surpluses cannot be evened out.)

14.4.5 Studies of actual market behaviours

This brief introduction to the price mechanism cannot include examples of the increasing number of detailed studies. The variety of technological and economic settings is very great. Some further readings are suggested below. Here, however, three general pointers are useful:

(1) Experience shows that markets are often wider and more dynamic and competitive than was first expected from apparent imperfect competition in parts of their structure.

(2) Monopsonies and monopolies can usually be established only by governments, and often have long term results very different from those intended.

(3) Many obstacles within international markets are commonly due to locations of skills, capital and infrastructure, and consumer incomes, and to tariffs and quotas, rather than due simply to numbers and sizes of firms as affecting pricing behaviour and restrictions.

14.5   INFLATION AND DEFLATION

In this review of marketing, prices have so far been considered as unaffected by general inflation or deflation. Prices have been in 'real' terms. This is because the supply and demand schedules that affect 'real' prices are for many detailed purposes better considered separately from the *general* upward shifts of these schedules when, with inflation, money falls in value; and from the downward shifts when, with deflation, money rises in value. Figure 14.3 shows how variable the general price levels in the USA and the UK have been. Figure 14.4 shows how much of the total variation in money prices of wheat in the UK is explainable by the variation in the UK's general price level, and how much remains in 'real' prices to be explained by changes in the particular supply and demand schedules for wheat itself.

General price levels change when the supplies and demands for money itself change. This is most obvious when a government spends heavily, borrows heavily, prints more money, and causes loss of confidence in its money. It thus

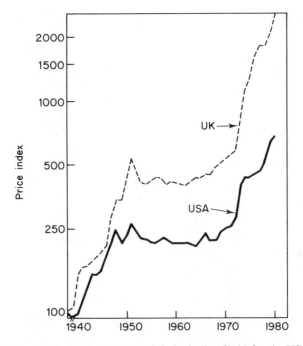

Figure 14.3   General price levels (raw materials, including fuels) for the USA and the UK
(1938 = 100).

lowers the foreign exchange value of the money because the supply of it is
increased and the demand reduced. Basic prices in its terms rise above the
international level that they otherwise would have followed if a constant
exchange rate had been maintained. But this international level (measured in,
say, US dollars) is itself unstable, for many monetary and related public finance
reasons (Figure 14.3). These are beyond the scope of this book.

We should note, however, that during periods of inflation or deflation different
types of price and marketing margin commonly rise or fall at different paces. This
was particularly important for farm firms during *past deflation periods* because
prices paid to them for farm products fell faster than urban wages, interest
charges on old loans, and many transport and other marketing costs. Many
sellers of labour and other services were able to maintain their 'prices'. In periods
of reduce demand, mining and manufacturing firms reduced production and
labour and other inputs, and tended to keep prices up. Governments maintained
the urban unemployed. Agriculture could not follow in the same way without
greater direct cost, because of its biological nature and because of the direct
responsibilities of its many small firms for individual households. Agricultural
production was therefore generally maintained and agricultural prices fell more.
In *past inflation periods* the 'terms of trade' became favourable to farm firms
(Table 14.2).

In *recent decades* many urban wage rates and other institutionally negotiated

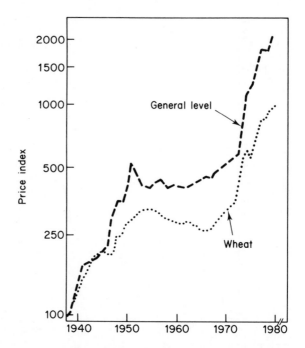

Figure 14.4   General price level, and prices paid to farmers for wheat, UK (1938 = 100).

prices affecting marketing costs and retail prices, and the prices of oil, machinery, fertilizers, and other farm inputs, have risen fast with 'inflation'. But many governments have not allowed previously protected agricultural product prices to rise so much. The money price ratios that have resulted are similar to those in past deflation periods – foodstuffs and raw material prices low relative to urban wages, interest rates, taxes, and marketing margins. Thus although the general level of prices has risen, there is some stagnation of economies – 'stagflation' (Table 14.2). And some countries in the Tropics and Sub-Tropics have relied heavily on cheap grain imports (e.g. Iran, Pakistan).

The economics of the individual farm firm greatly depend on the ratios between prices. These guide, as we have seen, many decisions about what and how much to produce, how, where, and when; and they greatly affect net incomes, savings, and investments. In the wider and more complex national and international economies price ratios likewise greatly affect millions of decisions that are together vitally important for economic welfare. When inflation and deflation result in big, general changes in these ratios serious socioeconomic tensions are liable to result between groups whose products become 'too cheap' and those whose products or services become 'too dear'. Such tensions were great during the 1920s and 1930s (e.g. within the USA and Germany, and between the less developed and the industrialized economies). Since 1974, tensions have increased again because of the changes in price ratios resulting largely from (i) the

Table 14.2 Index numbers of prices paid to farmers for their products, in money and real terms, in the USA.

	Deflation		Inflation		Stagflation	
	1926 to 1937	1937 to 1939	1932 to 1937	1939 to 1950	1966 to 1971	1971 to 1980
			number			
Years in period	6	2	5	11	5	9
		index numbers in last year of each period (first year of each period = 100)				
Money terms	71	78	188	272	106	219
Real terms in relation to Farm inputs						
Machinery	73	78	182	151	83	82
Fertilizers	86	80	211	189	118	82
All goods	82	87	169	135	94	89
Interest rates	88	88	298	316	63	51
Taxes per hectare	76	77	255	155	89	139
Family living costs	88	83	188	133	88	107

Source: Calculated from prices received and paid by farmers in the US Department of Agriculture's *Agricultural Statistics* for various years

rapid raising of export prices for petroleum, (ii) monetary and public finance policies following the concentration of much more purchasing power in the petroleum-exporting nations, and (iii) the reactions of those negotiating urban wages. Table 14.2 shows how the real value of farm products to farmers changed in periods of deflation, inflation, and 'stagflation'.

REFERENCE

Barlow, C. (1978). *The Natural Rubber Industry*, Oxford University Press, Kuala Lumpur, pp. 320–321.

FURTHER READING

Abbott, J. C., and Makeham, J. P. (1979). *Agricultural Economics and Marketing in the Tropics*, Longman, London.
*Barker, J. W. (1981). *Agricultural Marketing*, Oxford University Press, London.
Halcrow, H. G. (1980). *Economics of Agriculture*, McGraw-Hill, New York.
†Handy, C. R., and Padberg, D. I. (1971). A model of competition behaviour in food industries. *American Journal of Agricultural Economics*, 53, 182–190.
Kohls, R. L., and Downey, W. D. (1972). *Marketing of Agricultural Products*, 4th edn, Macmillan, New York.
†Marion, B. W., Mueller, W. F., Cotteril, R. W., Geithman, F. E., and Schmelzer, J. R. (1979). *The Food Retailing Industry*, Praegar, New York.

210

†Martin, L. R. (ed.) (1977). *Survey of Agricultural Economics Literature*, Vol. 1. North Central Publishing (for the American Agricultural Economics Association), St Paul; see chapters by B. C. French, D. G. Johnson, W. G. Tomek, and K. L. Robinson.

Mortenson, W. P. (1977). *Modern Marketing of Farm Products*, 3rd edn, Interstate Printers and Publishers, Danville, Ill.

Rhodes, V. J. (1978). *The Agricultural Marketing System*, Grid Publishing, Columbus, Ohio.

*Ritson, C. (1978). *Agricultural Economics: Principles and Policy*, Crosby, Lockwood, Staples, London.

Shepherd, G. S., Futrell, G. A., and Strain, J. R. (1975). *Marketing Agricutural Products*, Iowa State University Press, Ames, Iowa.

US Department of Agriculture, Economic Research Service (periodic). *Agricultural Outlook* publications, USDA, Washington, DC.

University of Saskatchewan (1981). *Guide to Farm Practice*, University of Saskatchewan, Saskatoon; see section on marketing.

Whetham, E. H. (1972). *Agricultural Marketing in Africa*, Oxford University Press, London.

†Tomek, N. G., and Robertson, K. L. (1981). *Agricultural Product Prices*, 2nd edn, Cornell University Press, Ithaca, NY.

QUESTIONS AND EXERCISES

1. Define a local market, and the larger markets of which it is a part.
2. Select a particular farm product. Indicate and briefly explain what in its markets cause consumption, in and out of store movements, and 'imports' and 'exports' to respond to trends and fluctuations in production.
3. List and briefly describe the physical movements and processes in marketing one farm product from farms to consumers.
4. In the nation you know best what are the ways by which government affects the prices of one particular farm product? What are the policy goals?
5. What are the determinants of changes in the quantities sold of one important farm product from farms in the region you know best?
6. Briefly describe one example of imperfect competition in the marketing of a farm product from farms to consumers.

Chapter 15

Factor markets I: Land and land use

15.1 INTRODUCTION

As human population pressures rise, and some uses of land are directed to producing crops or animals for sale, land values rise. These values are expressed as *prices* when rights to use, lease, and sell are transferred from one owner to another, or as *rents* when only the rights to use are transferred to tenants for periods under agreed or assumed conditions. Thus we conceive 'markets' in land and land-use rights, with buyers and sellers.

The ownership and use of land are important in social structures. The status of tenants is commonly, but not everywhere, lower than that of owners, who have different social roles and responsibilities, and greater power. The landless in the poorer agrarian economies have the lowest status.

For whole societies the distribution of rights in land is intended to be advantageous in maintaining territorial integrity and social cohesion, in securing foodstuffs, fuel, and other biological products, and in conserving resources for the future. Customs and legal codes therefore have long histories of elaboration in different societies. 'Markets' for land and land-use rights cannot be well understood without recognition of the importance of social values and goals, customs and laws.

Before setting out the whole range of functions in 'land tenure', we should recall, from the previous parts of this book, major determinants of the supply and quality of land for particular purposes, and also major determinants of demand (Chapters 3, 4, 9, 12). In particular, we should note again how inescapable is the use of billions of sites, and how variable sites are in natural endowments, man-made investments (and dis-investments), responses to further inputs, work requirements, locations as affecting transport costs, and reliabilities of yields from year to year. Also how much variation there is in sizes and organizations of farms as affecting sites.

15.2 BASIC FUNCTIONS OF LAND TENURE ARRANGEMENTS

Because the multitudes of local decisions that have to be made about agricultural production are best made mainly by individual farm firms (Section 12.6), land

211

tenure arrangements should help to provide the proper socioeconomic contexts for these firms. Items 1–11 of the list of basic functions (Table 15.1) are important for farm-economic sub-systems. Items 10 and 11 are also important for the distribution of income within socioeconomic systems; items 12–14 for this and other social reasons. Items 15–17 re-emphasize that all land tenure arrangements are subject to social values and therefore are controlled by customs and laws. The items most important for *efficiency* are 1–11; those for *equity*, 10 and 11 and commonly 14 and 16; those for *social cohesion*, 10–17. The scope for differences in political emphasis as between these three groups is obviously wide.

The following paragraphs explain briefly the functions, common ways of trying to ensure that they are satisfactorily carried out, and common difficulties and conflicts.

15.2.1 Distribution of land-use rights to secure the greatest economic production

The *average* gross outputs, net outputs, and 'profits' per hectare of land vary

Table 15.1   Basic functions of land tenure arrangements.

1. Ensure that those who will economically obtain the greatest production from land have use of sufficient land
2. Conserve fertility of land
3. Conserve other natural resources – that is, secure economic distribution of their use over time
4. Encourage investment in land improvements, where economic; and therefore provide security of tenure and compensation for improvements
5. Encourage use of outside capital where such use is economic; but safeguard against excessive borrowing, and expensive, inflexible credit arrangements
6. Provide flexibility in sizes of farms occupied by individual firms at different stages in their life, and with different degrees of initiative; and therefore provide mobility from farm to farm
7. Maintain, as far as possible, economic sizes of farms
8. Secure and maintain economic layouts
9. Provide safeguards against undue inflation of land values
10. Prevent undue inequalities in distribution of rights in property or in distribution of income within nations or econoic regions
11. Where land is leased to tenants, ensure that rents are appropriate in amount and form
12. Ensure that occupiers are 'socially compatible' with the communities in which they live; promote social cohesion
13. Avoid aggravating conflicts over religion
14. Avoid inappropriate use of ownership or occupancy of land as basis for distributing rights to vote, or other powers or responsibilities within social structures
15. Provide for acquisition of land by government and other bodies for purposes that are judged socially worthwhile
16. Facilitate taxation for finance of local and central government services
17. Be well understood both as to 'custom' and 'law'; provide adequate and cheap machinery for settlement of disputes over land

widely amongst individual farm firms, both within and between local areas (Chapters 11 and 12). The *marginal* outputs and 'profits' also vary, so that to shift units of land from the less efficient to the more efficient users would increase overall production economically. Maximum overall production would be achieved if land-use rights were so redistributed that any further *marginal* additions would have the same net outputs per hectare no matter to which user they were made.

For efficient overall production this ideal is of fundamental importance. In practice, it is difficult to achieve because (i) the many variables in farm-economic sub-systems and related biological and work sub-systems complicate assessments of land and of farmers; (ii) land itself is immobile (barring wind and water erosion) and so redistribution requires mobility of farm firms themselves, but this is commonly retarded by many socioeconomic 'ties', as well as by uncertainties.

These difficulties are substantially reduced if a market for land-use rights operates to show their 'prices' – i.e. the net value products (outputs less estimated inputs and required rewards) that buyers expect them to produce, and alternative use values that sellers have in mind. These are the best indicators of (i) the complex pattern of marginal products and (ii) changes over time. Annual contractural cash rents can be such 'prices', but they are affected by customs and laws. Prices for larger 'bundles of rights' in land such as owner-occupancy rights in the USA, and the UK, are affected by demands for land for holding (e.g. as *real* property instead of money funds during periods of inflation), or for other non-agricultural purposes. Such prices may differ widely from current marginal products in agricultural use. Therefore the various determinants of supply and demand in markets for land, and tenancy conditions as affecting markets for land use, may be such that achievement of the most economic distribution of land-use rights for efficiency in farm production is not fully assisted by 'pricing'. But the mathematical complexities of the distribution problem are such that no good solution without the use of prices is conceivable.

Even in societies with no serious disagreements about wealth and income distribution, the actual distribution of land-use rights is commonly widely dispersed and skewed, with many firms using small areas; and few using large areas (Table 12.8). The determinants of the distribution are basically the biological, work, and farm-economic variables, as managed by the 'human' variables in farm firms. Conflicts about distribution for equity, and social cohesion, as against distribution for efficiency in production, arise when societies begin to judge that the social status and economic power of those owning large land areas are excessive in relation to the status and income of small tenants or the landless. Taxation arrangements may not have maintained socially acceptable wealth and income distributions. Changing technological and market opportunities may favour the farms with more land, and they may be able to grasp these opportunities quicker. Such are the conditions that lead to demands for 'land reform'. But nowhere is maximum efficiency in production likely ever to be possible if all farm firms use the same amounts of land.

15.2.2 Conservation of fertility and natural resources, and encouragement of investments in land improvements

These functions (items 2, 3, and 4 of Table 15.1) are considered together because they are all related to future streams of inputs into and outputs from land. We have noted in Chapters 3, 4, and 5 how soil fertility can be improved or depleted by decisions about soil management, cropping, and animal management. We noted the 'biological' improvements that could be made by flood control, drainage, soil and water erosion controls, irrigation, glasshouses, grazing controls, buildings for animals, etc. We noted the biological importance of water conservation and shelter and therefore of trees. Also the improvements for 'work' that can be made by careful layouts (Section 8.5). All these require investment of assets, or foregoing of some output, so that, during a later period, the stream of outputs will be more favourable relative to the stream of inputs than it otherwise would have been. The farm-economic assessment of alternatives can be by the *present value method* (Section 12.4.3). An example of possible investments in water erosion control for soil and water conservation illustrates basic principles. Total inputs in the first year will be high and some types of further input will be higher later than in plans without these erosion control practices (e.g. inputs for maintenance of contour bunds). Outputs in the first few years will probably be lower, but outputs later may be higher; and total inputs other than for erosion control may well be lower *relative to outputs*. Assets in land at the end of say 12 years may be much more productive and valuable than in plans without conservation practices. Much depends on expectations about future prices of outputs and inputs, and on the discount rate used.

Experts and officials employed by governments may calculate the present values of conservation plans very differently from individual farm firms. Officials may expect the plans to increase future outputs more, and to reduce more the need for future inputs relative to outputs. They may expect prices of future outputs to rise more due to scarcities, and prices of future inputs to fall less (e.g. due to population increases and unemployment). They may use a zero or low discount rate (e.g. because they value net incomes in the future as highly as those in the present). But low-income families, with short lives and many uncertainties in unstable societies, may have effective discount rates of 25–40 per cent or more.

Similarly, landowners and tenants may have differences of opinion if they have different expectations about output–input relations and prices of outputs and inputs, and have different effective discount rates and 'time horizons'.

These 'time horizons' are determined by customs and contracts as well as by interest rates. Some tenancies are for only one growing season. Many do not provide full security to tenants for longer periods. Many do not provide compensation to tenants for the asset values of improvements remaining at the ends of tenancies. Many do not provide for inheritance of tenancies so that heirs can enjoy the benefits of improvements. But substantial depletions of fertility can be caused by some tenants who later move elsewhere without full compensation to owners.

Substantial depletions of timber, water, and other natural resources together with substantial pollution can take place if the consequences are 'external' to the economies of individual firms (e.g. when wood is taken from government-owned or common land; when excess nitrogen, faeces, and urine are leached into water courses).

15.2.3 Encouragement of economic use of outside capital, but avoidance of faulty financial arrangements

Assets in land are physically fixed (bar erosions or depletions) and their values are more stable than values of other assets, or net revenues. They are therefore regarded as good security for lenders and in most countries many loans are made against them. Both borrowers and lenders can, however, make mistakes (e.g. when loans are excessive in relation to the total assets, net worths, and the levels or variations over time in productivities and net revenues; or, on the other hand, when borrowers voluntarily ration their total assets at uneconomically low levels for fear of borrowing) (see Sections 11.5 and 18.2.2). Where the 'profits' of farm firms are low and highly variable and where the finance of owner-occupiers is not well and widely understood, it is particularly desirable that land ownership is neither misused as a basis for loans, nor under used.

Assets financed by landowners, using land as security, can be used by tenants in return for rent payments, but the assets that tenants themselves can offer as security are commonly inferior. Where total asset requirements are high this is especially important in determining which assets should be provided by landowners and which by tenants, as well as the types and conditions of loans to tenants (see Section 18.2).

15.2.4 Provision of flexibility in farm sizes

The abilities and efforts of farm firms tend to change as their lives progress. This is due partly to natural changes in physical energies and ambitions as individuals and families mature and age, partly to accumulation of knowledge and experience, and partly due to the tendency for net worths to be low early in firms' lives and then increase. Therefore, in addition to the variations in farm sizes needed for our first function, some changes in sizes are desirable to suit the demands and productivities of firms at different stages. Where land is available for cropping at little more than the cost of clearing vegetation, young men can (e.g. in parts of Africa) secure use of tillable land by themselves making the investment of labour for clearing. Where land is scarce and highly priced, some expanding firms can rent some additional parcels or farms to add to the land they already use. And additional land may be purchased, perhaps incurring additional liabilities for some of the cost. In the USA, firms that both own and rent farm land use as much as 53 per cent of the total land in farms. But even so, social ties and institutional arrangements tend to retard the mobility of land-use rights between users, so that differences in marginal productivity persist.

15.2.5 The maintenance of economic sizes of farms

In any one country the average size of farms as measured by land area is determined by (i) the total land area available; (ii) the area of land not taken into farms; (iii) the total population; (iv) the population employed other than in or by farm firms; (v) the total population employed in or by farm firms per firm. These determinants are the outcomes of complex demographic, political, social, and economic forces. Thus Indonesia is a country with the sixth largest human population in the world and a total land area of 1.2 hectares per person. But much of this land has infertile soil, and the high social values of living close to family and ancestors have also retarded migrations and the 'bringing in' of more land. Although urbanization and industrialization are proceeding, 80 per cent of the total population is still rural and a high proportion of these people have farms. Therefore the average size of farms is small (Table 12.8).

The distribution of farm sizes about the average in any country has been determined basically by the biological, work, and farm-economic variables and the development and distribution of abilities to manage them. These and other forces have contributed to (i) the historical determination of the social structure, with land tenure groups each with particular responsibilities and powers, roles, and status; (ii) the development of financial, marketing, and other services to farm firms; (iii) customs and laws affecting inheritance of rights in land; (iv) taxation of incomes, capital gains, and capital transfers; (v) other laws and customs affecting competition between land users and owners, and transfers of rights in land.

Thus, in most of Great Britain, the feudal social structure and the custom of bequeathing land so far as possible only to eldest sons led to large estates, with much of the land let in farms to tenants. The farms were varied in size partly to make the total rental values higher. In the twentieth century (ii)–(v) have all changed, so that now more farm firms own their farms and fewer rent them, and the dispersion of sizes has changed but is still wide (Table 12.8). Since the eighteenth century in Northern Ireland, many of the determinants of average size and of the dispersion of sizes have been different. Within the USA and Canada, the determinants in different regions have also differed.

Many societies attempt to define the 'minimum economic size' of farms and even take administrative measures aimed at preventing farms from being 'too small'. But the complex of determinants of actual sizes (including differences in land qualities and the qualities of managers) commonly defeats such attempts. This complex also alters within a few years the uniformity of sizes often planned for official land settlement schemes.

15.2.6 Securing and maintaining economic layouts

Holdings became fragmented where, in allocating land to individual farmers, attempts were made to give each a share of different types of land, and where inheritance customs shared equally between sons (and perhaps between daughters). Thus even today in many countries, scattered, small, badly shaped fields remain that are awkward and costly to use. Consolidation of holdings is inevitably difficult and slow, and usually cannot proceed without the support of special laws (e.g. in western Europe).

The re-laying out of farms for easier and lower cost use of machinery and higher labour efficiency (see Section 8.5) can proceed most readily on consolidated owner-occupied holdings. But provided that landowners and tenants understand the costs and the benefits, and can agree appropriate changes and sharing, progress can be rapid also on tenanted farms.

When planning changes both owners and tenants should take into account water control, and water and soil conservation, and also conservation of wildlife and landscape (Section 8.5).

15.2.7 Safeguarding against undue inflation of land values

Rising population pressures, more commercial crop and animal production, urban developments, and expansion of non-agricultural uses – all these increase land values. Increases in prices and rents help to allocate sites to appropriate uses, even although arguments arise about who should gain directly from the increases themselves. But land values also rise during inflation periods and when demand is temporarily at exceptionally high levels (e.g. on return of soldiers entitled to 'resettlement' by governments; or when other investments are unattractive). In some circumstances therefore land values can become, for a time, misleadingly high. Fuller economic analysis and the spread of information and understanding are the best safeguards. These should be made easier by simple registration of ownership and tenancies, and reporting of prices and rents. Occasionally governments may feel it desirable to impose price and rent controls for a period of years (e.g. in New Zealand in 1946).

15.2.8 Prevention of undue inequalities in distribution of rights in land

We have recognized that the distribution of land-use rights that would make the *marginal* value product of land the same for all users would maximize the total net agricultural product, but would leave the total products per firm and per hectare to vary widely (Section 15.2.1). With actual land-use distributions, they do so vary (Tables 11.1 and 12.5). The major reasons for social demands for 'land reform' are wide differences in land-use rights and incomes between tenure groups of different social status, from the landless to large estate owners (e.g. in Latin America). But 'land reform' plans inevitably have effects on the whole range of land tenure functions. The actual and probable consequences of plans therefore require close study and widespread understanding. In many countries, adequate plans now seem essential to any major increase in agriculture production. But everywhere, narrow, inadequate plans can raise welfare in a patchy and ill-balanced way, or even reduce overall welfare.

The most widespread land reforms of the twentieth century were carried through in Mexico and by communist governments in the USSR, eastern Europe, and China. The experiences gained deserve comprehensive study beyond the scope of this book. But we should note that the results are complex and varied. They are now widely recognized as requiring new reforms both for efficiency, and for equity and social cohesion (see also Section 23.5.3). Collectivization of farms does, of course, pose questions about management control and, perhaps even more important, about motivations and rewards for labour. These are introduced in Section 16.3.1.

Within nations, inequalities between regions are common because farm families have not shifted away fast enough from areas with natural endowments and geographical locations unsuited to the price and wage ratios of recent times. (For example, the small farms of wet acid soils in the outer islands of north-western Europe cannot provide incomes that will keep young men and women from emigration elsewhere unless they have other sources of incomes that are higher than in the past. Much emigration has taken place, and government subsidies are high for social reasons in 'less favoured areas'. But so long as national levels of income per head are far above the incomes possible from small poor farms, disparities will probably remain. How high subsidies should be must depend on social judgements. And this is true also of the balance of purposes in land reform between (i) retaining old socioeconomic distributions of land-use rights and (ii) securing greater output per hour of labour in agriculture in the future by redistribution).

Nations jealously guard their powers of control of immigration. So historical determinants of population pressures are still evident in the wide differences in land resources per head in different countries. Modern technology has provided opportunities to reduce the inequalities by raising crop yields. International trade in agricultural and other goods and services provides other opportunities. So long as national societies are not prepared to accept much greater immigration and related cultural changes, international disparities in income and wealth can be reduced only by more widespread use of technology, and more trade.

15.2.9 Ensuring that rents are appropriate in amount and form

Contractural rents are prices paid for land use and have as their main function the matching of supplies and demands in the complex pattern that will so distribute land use between users that its total product is maximized. Coming as near as possible to this ideal has a high social value. But progress towards it is difficult to measure. Societies and governments are often more impressed by how unequal incomes are. And they wish to reduce the incomes of landowners and increase those of tenants by keeping rents below their economic equilibrium levels. Societies as well as individual tenants and owners also wish to have some control of how the consequences of year-to year variations in crop yields and of price fluctuations are shared between owners and tenants. Therefore the *form* of contractural rents is important.

Attempts to keep rents down are usually effective only if they are indirect (e.g. by redistribution of income and capital by taxation; by 'bringing in' more land; in Great Britain, by giving existing tenants and their heirs complete security of tenure and allowing rent increases only by mutual agreement or after arbitration).

The possible forms of rent are as follows:

(a) *Share rents* Agreed physical shares of the annual output of the main product(s), so that yield and price variations are shared between owner and tenant. The owner may provide agreed shares of fertilizer and other input costs, and some management. If there is not appropriate sharing of inputs, intensification may be unattractive to tenants because their rewards for marginal costs would be only their *shares* of marginal products.

(b) *Produce rents* Agreed physical amounts, so that owners bear the uncertainties of prices, and any costs of marketing.

(c) *Tribute rents* The annual delivery quotas of products required of collective farms by communist governments may be regarded as rents because they are in part essentially annual payments for the use of nationalized assets. Although they are paid for, their prices are well below what market prices would be.

(d) *Cash rents* Contractural annual payments for the use of the assets provided by owners (commonly land, buildings, and other fixed equipment). The tenant is the farm firm with responsibility for most economic decisions and carrying most risk and uncertainties. Ideally, owners have rights to compensation for any depletion of their assets, but have to pay compensation for unused fertilizer nutrient residues and other useful improvements left by tenants.

Cash rents are most suitable where land values are high, much other capital is required for economic production, the physical sharing of production would be too complex to manage, and credit and taxation arrangements are well enough developed to reduce the problems in year to year variations. But laws and regulations about cash rents may give excess priority to some social goals as against others (e.g. aggravate immobilities of land between users).

Share rents are suitable where sharing of produce, inputs, and management can be well arranged and intensification usefully fostered by owners' leadership, as well as by the sharing of risk and uncertainty bearing.

Produce rents are only suited to some simple farming systems.

Tribute rents secure deliveries to government agencies of collective farm products so far as possible in accordance with central plans, but they tend to have the disadvantages that (i) they aggravate year to year problems of the collectives as farm firms, and (ii) they do not well measure potential marginal productivities of land, because collectives do not bid for land transfers by offering rents or other '*prices*'.

15.2.10 Securing social compatability and cohesion in rural communities and avoiding conflicts over religions

The occupation and ownership of land are the natural territorial bases for continuity of 'folk ways' and mores, and ideas about social structures and functions. Strangers with different cultures, if they come in substantial numbers, are often not compatible, and so social cohesion is reduced. Even in the settlement of sparsely populated areas, social conflicts can result if the settlers are from very different cultures. Land tenure arrangements should help to solve the problems rather than aggravate them.

15.2.11 Avoiding inappropriate use of land rights to define political and social structures

Social structures and definition of each status within them (including rights to vote and control governments) are in agrarian countries commonly closely related to distribution of land ownership and land-use rights. This is so both where these are comparatively equitable and where they are not. We may well ask therefore whether, as a result, opportunities to improve agriculture and welfare are grasped earlier or changes of many kinds are delayed. But the timings of changes depend on the orientations of societies' values and goals, and on how much social cohesion has been achieved and maintained. These are determined by processes as well as structures. Generalizations are liable to mislead. Obviously, however, when non-agricultural activities develop, and government services become more complex, social structures should alter, because otherwise many political and administrative decisions will not be appropriate. This was shown by agrarian structures that attempted in the nineteenth century to govern economies that were being industrialized and urbanized.

15.2.12 Providing release of land to government agencies for social purposes

All societies have well established rights to acquire land for social purposes that are judged more important than those of individual firms. In England and the USA, an important right is that of *eminent domain*. Thus for roads, rail roads, airfields, defence, and the like, land can be acquired. Compensation related to market rates is usual. This right, together with laws and regulations that 'plan' or 'zone' land-use categories, can determine what is and what is not land for agriculture.

15.2.13 Facilitating taxation

Land ownership and land use provide bases for taxation which are especially important in economies where some other types of tax (e.g. income tax) are difficult and costly to collect (Section 19.4.5).

15.2.14   Being well understood and providing for arbitrations

For both economic and social reasons land tenure customs and laws, and changes in them, should be widely understood. Transfers should be easy and cheaply made and any local and personal disputes should be wisely, consistently, easily, quickly, and cheaply settled. Registration of titles by modern methods is therefore desirable.

15.3   STATE OWNERSHIP

Governments commonly hold the main rights in land that has not been settled or is very sparsely populated. They can more readily control settlement, conservation, and new uses (e.g. in much of the western USA and Alaska). They can also secure government revenues for sales of rights and so prevent early private owners from acquiring all the increment in land values due to improved transport and other infrastructure and increased demands for land.

State ownership is often proposed as a means of ensuring that other functions in land tenure arrangements will be better carried out (e.g. securing greater intensities in use; favouring better sizes of farms and layouts and more fixed investments; having flexible rents; avoiding inequalities of wealth and income). Two basic questions arise however, in every country:

(a)   Would the government organization and management actually make well the multitudes of detailed, economic, and social decisions required, so that *all* the functions of land tenure would be carried out with priorities related to real values and goals? Would a tenants' union become too strong a countervailing economic and political power?

(b)   Would the new responsibilities and powers of government members and officials (their new status) continue to be socially acceptable, and help to secure and maintain social cohesion?

As indicated in Section 15.2.1, 'prices' are needed to solve the problems in the allocation of land resources. Most governments have concluded in the past that such 'prices' are best determined in competitive markets rather than under an overall government monopoly. They have in many countries amended the long-run effect of the market system by policies on taxation, sizes of land holdings, conditions of tenancies, agricultural loans, subsidies, conservation, zoning and planning, etc. But ideal fulfilment of all the functions of land tenure systems is nowhere yet achieved.

FURTHER READING

†Bertrand, A. L., and Corty, F. L. (eds) (1962). *Rural Land Tenure in the United States*, Louisiana State University Press, Baton Rouge.
*Food and Agriculture Organization (1979). *Agrarian Reform and Rural Development: National and International Issues for Discussion*, FAO, Rome.

*Fones-Sundell, M. (1980). *Agrarian Reform and Rural Development: Theory and Practice*, International Development Centre, University of Uppsala, Uppsala.

Halcrow, H. G. (1980). *Economics of Agriculture*, McGraw-Hill, New York.

*King, R. (1978). *Land Reform: A World Survey*, G. Bell, London.

Lim, S.-C. (1980). *Marketing Rubber by Smallholders in Peninsular Malaysia*, Malaysian Rubber Research and Development Board, Kuala Lumpur.

*Peters, G. H. (1970). Land use studies in Britain: a review of literature. *Journal of Agricultural Economics*, **21**, 171–214.

Reiss, F. J., Ballard, E. M., Brink, W. H., and Harryman, W. R. (1980). *Farm lease practices in three South Central Illinois counties in 1980*. Report no. AE-4491, Department of Agricultural Economics, University of Illinois, Urbana-Champaign.

US Department of Agriculture (periodic). *Farm Tenure and Cash Rents in the United States*, US Government Printing Office, Washington, DC.

Warriner, D. (1969). *Land Reform in Principle and Practice*, Clarendon Press, Oxford.

QUESTIONS AND EXERCISES

1. Briefly define the farm land tenure arrangements that you know best. Which of the functions set out in Section 15.2 do you think are satisfactorily carried out; and which are not?

2. What proposals are made for improvements in one particular set of farm land tenure arrangements? Indicate which other functions of land tenure might be less well carried out if the proposals were put into effect.

Chapter 16

Factor markets II: Labour and management

16.1 INTRODUCTION

16.1.1 Basic biological and work determinants

In Parts II and III, we noted the following characteristics of tasks in crop production:

(i)   How closely their type, details, and timing should be related to biological conditions on each individual site
(ii)   How variable these conditions are from site to site, even within individual farms
(iii) How short are the periods during which the many tasks in crop establishment, weeding, pest control, and harvest can best be done, and how variable these periods can be from year to year
(iv)   The wide range of equipment that can be combined with labour in variable ratios to accomplish tasks, and the trends in developed countries towards more complex equipment and skilled labour, and less human physical work; but, in other countries, the heavy human work that is still required with its burden of physical stress and disutilities.

We also noted the following characteristics of tasks in animal production:

(i)   How closely they should be related in detail and timing to the varying biological needs of individual animals (e.g. to secure high reproductive performances)
(ii)   How the time and work for whole herds or flocks depends on production systems and labour–equipment plans and on observations, many skills, and prompt flexible adjustments (e.g. to prevent disease)
(iii) How high productivity of grasslands depends on skilled, timely management.

Together these characteristics make high biological and farm-economic efficiency in agriculture impossible anywhere unless (a) management has much

223

knowledge of biological conditions site by site, and animal by animal (e.g. in dairy and pig production) or group by group (e.g. in sheep and poultry production) and keeps this knowledge up to date; (b) management and labour are therefore observant and able and willing to take prompt, skilled action; (c) management and labour are able and willing to complete tasks within limited time periods, despite personal physical stresses and the temporary loss of leisure.

These three needs have determined the organization of agriculture as it is in most of the world. In all regions multitudes of firms each with their own sites and animals have small management and labour forces with intimate local knowledge, a wide range of skills, and flexibility and persistence in the face of biological conditions and all that affects them. And persistence includes a willingness and ability to bear the biological and economic uncertainties of crop and animal yields and of prices.

Where larger firms are established the reasons are unusual (e.g. (i) plantations for tea and other specialist crops where the introduction of new production activities required many new skills, processing plants, and heavy capital investments, and therefore 'lumpy' organization and management such as only big units could afford; (ii) collective farms created on governments' insistence for particular goals, such as the procurement of foodstuffs at low prices for urban populations, and the redistribution of incomes and alterations in the power structure in rural areas). Experience shows that large firms have serious difficulties in trying to secure the three basic essentials by delegating many responsibilities to smaller units and attempting to provide them with the direct rewards and other motivations that are essential if these responsibilities are to be well fulfilled. In general, large plantations require especially capable management and well designed piece-rate and other wage payment systems. Collective farms require particularly strong social cohesions amongst individual families (e.g. as in the early *kibbutzim* of Israel and Huttite and other religious groups in the USA and Canada).

16.1.2 Technological changes

Actual and potential technological changes add to the functions of management (e.g. in revising decisions about choices of product and input mixes) and of labour (e.g. in acquiring new skills in machinery operations and servicing). The will and ability to acquire new knowledge, and apply it appropriately on particular sites and in particular herds or flocks, thus become increasingly important. They determine the productivity of management and labour in each firm, and so are the major determinants of the rates at which individual firms 'grow' or 'decline' (Tables 12.5, 12.6, and 18.1).

16.1.3 Structural changes

The development of socioeconomic systems includes (i) shifts in the geographical pattern of production; (ii) increasing dependence of agriculture on purchases of factors and sales of products; (iii) reductions in agriculture's share of the total labour force; (iv) some changes in sizes of farm firms; (v) increases in taxation to pay for many government services.

These changes are still being made even in 'developed' economies. They all require additional management skills and energies (e.g. in changing choices of product, in marketing, in management, and in accounting, finance, and taxation matters).

16.1.4 Social changes

In self-sufficient local economies, before marketings of produce for money have developed, production can be managed largely by a patriarchal structure of large extended families. But when cash rewards for extra labour can be secured directly (e.g. by young men from sales of produce), such structures begin to lose control. Few areas now have large extended families controlling management and labour.

History and studies of present day primitive cultures also show that, early in development, societies distinguish different types of task, and different social groups responsible for them. The grouping is by sex, age, parentage, and other determinants of social status. Even slavery can occur early if there is local shortage of land, or war about territorial boundaries (e.g. early in civilization in the Middle East, and in Ancient Greece and Rome). The feudal social systems of Europe had well defined 'levels' of responsibility and authority. Commercialization and related or parallel social changes affect competition in all forms, and thereby regroup power and responsibilities. Commonly therefore, we now distinguish the following groups in relation to modern agriculture: the large, landless, urban populations; rural landowners who lease much land; owner-occupiers; tenants; family workers closely related to heads of farm firms; regular hired workers; casual workers and the rural landless. All these may have agricultural functions as their only productive activities or they may be 'part-time'. Apart from land ownership, their agricultural functions may even be only seasonal. They may be regarded as skilled or unskilled; as 'specialized' or, more commonly, as 'general'. They may be local residents or migrants. They include children and even 70 year olds.

The economic and social relations between all these groups add further to management functions in farm firms (e.g. in determining conditions of employment and offers of pay) and to workers' problems (e.g. in seeking work and deciding what pay to accept). In other words, the market for labour is complex because workers vary widely in type and social expectations as well as in productivity. Many governments attempt to improve the employment conditions and pay of low income workers.

Farmers themselves must be regarded as part of the supply of labour and management, and net farm incomes include the rewards for their efforts as well as for the use of capital (Section 12.1.3). Many agricultural policies attempt to alter net farm incomes. And special taxation issues arise. So in these ways too governments are drawn to major questions about the supply of, and demand for, agricultural labour and management.

Questions inevitably arise about mobility within social structures, particularly mobility into the tenant farmer group from that of landless workers, and mobility within the farming group. The possibilities of mobility 'up' the social scale are sometimes called 'the agricultural ladder'.

Other important questions arise from (i) the development of contracting firms that sell to farm firms their services in mechanized field operations and other types of work; (ii) the comparatively small but significant socioeconomic group of hired managers.

16.2 MANAGEMENT

16.2.1 Demands

The demands for management are, like those for other factors of production, derived from demands for products. In agriculture, in addition to market demands, there are demands from farm households themselves for 'self-supplied' products (see Figure 13.1) which are still very important in many circumstances. The 'prices' paid for management are the components in net farm income other than returns for the farmers' manual work and for use of all assets (risk and uncertainty free) (Section 12.1.3). When returns to self-employed farmers' managements are compared, they can most usefully include those for risk and uncertainty bearing. But when returns to hired managements are compared they may not.

What 'prices' (net rewards) can be paid for management are determined ultimately by the productivities of management in satisfying locally expressed demands for products within the local constraints of factor availabilities and costs. The 'qualities' that determine productivities are therefore of fundamental importance. They include, as noted above: wide ranges of knowledge and skills related to the local biological, work, and farm-economic conditions; continuous observant care in updating knowledge of conditions; abilities to acquire useful new knowledge and skills for appropriate local use; reduction of risks and uncertainties by prompt action; willingness and ability to complete tasks efficiently within limited periods; effectiveness in family and other social relationships, including man-management; knowledge of taxation and other government measures; willingness to bear the risks and uncertainties that cannot be economically reduced, and farm-economic decisions (e.g. about liabilities in relation to net worth) that ensure survival. These and other 'qualities' are partly innate and partly the result of informal and formal types of education and training. Even within localities and societies with many educational and training opportunities widely available, individual managers vary much in their 'qualities' and therefore in their productivities and rewards (Table 12.6).

How closely the productivities and rewards of individuals are related to managerial qualities depends of course on whether individuals have opportunities to use land and other factors. Immobilities of factors between users and potential users may delay the expression of managerial qualities for one or more generations. In general, however, the distributions of farmers according to the 'sizes' of the input mixes that they have acquired and their rewards for management show wide variations, many of which are caused by actual managerial qualities. And it is significant that these distributions are skewed to the right (Table 12.8). To move up the ranges (i.e. to secure use of more resources and higher rewards) requires increasingly rare combinations of qualities of high order. Many more managers have qualities for the lower and left sides of the ranges.

Governments can affect the demands for management in various ways through (i) farm incomes (e.g. by subsidies, trade and other controls of product and factor prices, government purchase of products, or supply of factors); (ii) land reform, size-of-farm and rent policies; (iii) subsidies to secure early retirement, and amalgamations; (iv) taxation (e.g. on current incomes, and on capital transfers from fathers to sons); (v) cheap loans to help the first steps on the 'agricultural ladder'; (vi) promotion of particular production activities (e.g. cash crop production to secure earnings from exports); (vii) the complexity of government regulations and administrative measures.

Where governments make the levels of farm incomes favourable and reduce year to year variations in them, they alter the relative importances of various managerial qualities (e.g. of willingness to bear uncertainties) and therefore encourage shifts in the distribution of factors and rewards between managers. Government policies and market conditions may also cause changes in land ownership that lead to increased demands for hired managers (e.g. when leasing out of land becomes less attractive than using it directly with a hired manager).

The demands for hired managers are seldom assembled in an organized market. The market consists rather of a widespread network of personal contacts only some of which are based on advertisements. The sources of information on 'qualities' are largely personal and intuitive, although the results of formal education and training may in some circumstances be available and objectively used. The conditions that managers would work under are also often not well defined. Lack of information and poor information flows cause serious uncertainties. But immobilities of land away from management by heads of farm firms themselves are the main reasons for restrictions on demand for hired managers.

The demands for management by these heads of farm firms are expressed essentially through markets for land and land use, and through government policies. The position of a new tenant after open competition for a tenancy is the simplest example. The rent offered and accepted should be the lowest that was required to secure him the land-use rights. After deducting from the prospective gross output values (including non-material values) costs for other inputs (including use of his assets and his manual labour), there should be left the rewards that he expects for management (including risk and uncertainty bearing). In other words, the rent accepted after offer was an *indirect* and inverse measure of the demand for the tenant's management, and the price at which he would supply it. The market for land use determined whether demand for and supply of management were balanced at that rent. Where land is scarce (e.g. in much of India), manual labour is cheap, and management for traditional production is plentiful, rents are high, wages low, and management returns low. But where the technology of farming is highly developed and still changing (e.g. in central Illinois, USA, or central Norfolk in England), the best land is becoming more and more productive and labour and management are scarce, rents, wages, and managerial returns can all be high.

The rewards that owner-occupiers who have inherited their farms obtain from the socioeconomic system are more difficult to discern. They have not had to offer rentals. Their rewards have various components (Section 12.1.3). What the rewards for their management in an area have been can be judged only for a period of years in the past, using information about net farm incomes and their year to year variations; assets used; interest rates on risk-free uses of capital; changes in land and other values, and numbers of farmers, prices paid for land, and other indicators.

16.2.2 Supplies

Numbers of farms tend to change only slowly and are not good measures of the potential supplies of management that, with demand, determine the prices (rewards) of management. Numbers of potential managers can be considered as determined by the following: (i) numbers of children reared on farms; (ii) numbers of other children reared; (iii) numbers of youths wishing to enter into farm work and training with a view to eventually managing; (iv) numbers wishing to leave farm work before managing; (v) numbers never achieving adequate management 'qualities' (vi) migration inwards to a locality of managers, or workers who eventually could become managers; (vii) numbers wishing to leave management part-time before final retirement; (viii) desired ages at final retirement.

The basic determinants of these variables are many. In addition to all those of birth and death rates and the age structures of populations are all those of employment opportunities outside agricultural production. Different age groups (particularly those in rural areas) and different potential land tenure groups compare these opportunities with those in agriculture as regards prospective material incomes, non-material benefits, costs of living, security of employment and incomes, and conditions of work (including degrees of freedom from discipline). Costs of transfer may be considered. Information about all these varies in reliability and is not well distributed. Education and training tend, particularly in poor rural areas, to be inadequate for easy transfers to non-agricultural employment, but rural youths acquire much knowledge and skills informally about agriculture. Formal education and training for agriculture is well developed in some areas and absent in many others. Abilities to save so that farm assets can be acquired vary widely with managerial qualities and circumstances. Inheritance customs, taxation, and markets for loans and land and land use tend to determine those who aspire to own farms and who to be tenants. Retirement customs and social security arrangements for the old, and government subsidies, may also affect ages at retirement. The results of all these determinants naturally vary greatly between areas and countries and over time. In addition to differences in the *numbers* seeking to manage, there are also differences in managerial 'qualities'. Moreover, individuals also differ in what disutilities they feel when they make additional efforts, and in what satisfactions they gain from additional outputs. Age, health, education, and social experience all contribute to such differences, which are usually recognized as those in 'willingness to work' and 'demands for income'.

16.2.3 Prices (rewards) for management

Four generalizations of fundamental importance result from studies of demands and supplies of management in agricultural production:

(a) Different countries, and areas within countries, can have widely different demand and supply situations and therefore widely different levels of reward (Table 12.2).

(b) Almost everywhere the potential *numbers* of managers are large. The non-material benefits directly or indirectly from management are important. But the demands for numbers of managers (the number of farms) are inelastic. Therefore, to secure *numbers* of managers socioeconomic systems do not have to pay high material rewards.

(c) Governments can, to limited extents, raise farm incomes and reduce taxes, and so increase rewards for management. Governments can also affect the demand for managers through some alterations in the number of farms (e.g. increasing it by settlement schemes and land reform; reducing it by subsidizing amalgamations, and limiting the fluctuations in farm incomes). But commonly the changes in numbers are slow, and due more to changes in technology and qualities in management, causing amalgamations of farms.

(d) *'Qualities'* in management are very variable. Their importance increases as farm production technology becomes more complex. Education and relevant research and development can improve them, but do not necessarily make them more evenly distributed. Prospects of higher rewards can induce their supply (e.g. amongst self-employed farmers). Where sufficiently 'low-priced', qualities in management can induce some changes in numbers of farms (e.g. by inducing landowners to lease more farms or sell off some land, or by settlement of new lands).

(e) Variations in 'qualities' are amongst the major causes of wide variations in the total rewards of individual managers.

16.3 LABOUR

16.3.1 Demands and organization of markets

Demands for labour have their origins in the tasks in farm production systems. They are determined therefore by farm-economic decisions about choice of products (activities) and the 'input mixes' in activities. Thus demands for labour for irrigated rice or vegetable production with perhaps two crops per year are very different from those for sorghum and goat production in semi-arid conditions. Farm-economic decisions also determine how, when, and how well tasks are done, and therefore the ratios between inputs of labour and of equipment use, and also the types and 'qualities' of labour used. Thus demands for unskilled, migrant workers for seasonal harvesting of vegetables or fruit are high where mechanical methods with more skilled labour are not feasible or economic. But seasonal harvesting of wheat in the USA and Canada can be done by small amounts of regular farm labour, along with contractors supplying skilled labour and use of combine harvesters. The different circumstances in which farm-economic decisions are taken and the variations in the qualities of management are the basic causes of the very wide differences in demands for labour, and the different trends in these demands over time. Such differences need not be further elaborated here (Chapters 10–12).

For socioeconomic systems an especially important trend is that towards lower demands for labour as equipment designs are improved and substitution of more equipment use for labour becomes possible. In western Europe and North America such substitution has caused some 'push' of hired labour off farms, as well as making good the loss of labour due to the 'pull' of opportunities in non-agricultural employment. The trend has substantially reduced agricultural populations in rural areas. In some poorer countries the substitution possibilities have been quickly grasped by certain farmers (e.g. for mechanized grain production in Turkey), so that they have acquired more land and rendered others landless and unemployed. In many countries, however, interest in animal-drawn implements, other 'intermediate technology', and more intensive agronomic systems has increased, in the hope that demands for labour may be sustained and even increased, and total production raised further (Chapters 9 and 12).

Social values and goals also affect demands for labour in other ways. At the individual firm level many old and less able workers are still employed although their productivity is low (Table 12.1). In Tanzania and other parts of sub-Saharan Africa traditional 'working parties' are used for clearing of bush fallows and harvesting even though they can be inefficient. In Java there are social values in allocating out rice harvesting tasks very widely in traditional ways, although the costs are comparatively high. In some industrialized countries laws require that wages should not be below stated minimum levels, and that conditions of employment including safety precautions are regulated.

Even more important determinants of demands for *hired* labour are the numbers of farmers and their close relatives remaining on farms. Only the labour that they do not provide is hired from off the farms. Much therefore depends on how many farms result from the workings of markets for land and land use and what their distribution is according to total labour requirements. Thus in our Saskatchewan area, the total labour requirement is almost entirely provided by the farmers' own labour and the other demand is for only a little seasonal labour and contract work in harvest periods. Likewise there is little hired labour in central Illinois or on small-holdings in western Malaysia (Table 12.2). But in the Aberdeen area, the demand is for regular labour and some casual labour, making in all 63 per cent of the total, with the average farmer and his wife providing only some 37 per cent.

Where collective farms are established the demand for labour is based on membership of the collective and a variety of arrangements that (i) allocate individuals and tasks to brigades and teams; (ii) calculate the days worked according to complex 'weightings' of different types of task; (iii) calculate income and expenditure of the collective; (iv) divide the surplus according to days worked by individuals. Under some arrangements basic wages are paid, making smaller the surplus for later division. There is no substantial mobility of labour between established collectives responding to an informal market network. The rewards for labour differ widely between collectives. Migration away from collective farms to meet demands for labour in non-agricultural activities is commonly regulated.

An effect of farm numbers, sizes, and types of organization is thus that they determine how labour is rewarded, and this is of fundamental importance to both 'demand' and 'supply' in the labour market. Table 16.1 summarizes the main

Table 16.1   Summary of rewards for labour, by types of worker.

Types of worker	Money wage based on		Payments in kind	Surplus of firm
	Time rates	Piece rates		
Owner-occupier	...	...	...	All
Tenant				
Cash rent	...	...	...	All, net of cash rent
Share rent	...	...	...	Share
Member of farm household	a	a	a	a
Hired workers				
Regular[b]	+ + +	...	+	c
Part-time	+ + +	+ + +	+	...
Casual seasonal	+	+ + +	+	...
Collective farm member	d	e	+	Share according to 'days' worked

Codes are: + + +, main method or alternative main method; +, subsidiary method.

[a]Rewards range from those of regular hired workers to little more than subsistence, but usually include some training and may eventually include a share in the firm's net worth and ownership of the land or land-use rights.

[b]Year-round.

[c]Some payments may be 'by results', but in complex intensive farming systems these are difficult to calculate satisfactorily, so less precise 'bonus' payments are made.

[d]Some weekly payments may be made, reducing the shares.

[e]Some 'days' worked are defined as 'pieces' done.

ways. The rewards for labour of heads of firms (rows 1, 2 and 3) are not separate from the rewards for capital use, management, and risk and uncertainty bearings. But they are usually closely related to socioeconomic demands and to what individuals have provided. Thus between individual share croppers, rewards vary according to their individual efforts. Members of farm households other than heads of firms are rewarded in various ways, some of which may not measure demand, but cause uncertainties and poor supplies. Well thought out and agreed arrangements are desirable, especially between fathers and sons and daughters. Hired workers are offered definite payments in money or kind, or both, related to expected individual efforts, and management has responsibilities to ensure these are made. The rates are largely determined in local market networks. Neighbours and 'work parties' (as in Africa) usually expect payment mainly in reciprocal labour and in the form of food and drink. Some arrangements can be very efficient and fair (e.g. in sheep handling in the hills in north-western Europe). Some are much less so.

For the labour of individual collective farm members the rewards depend on individual efforts, but also on the management and resources of the particular farm and on brigade and team arrangements and efforts. There are no market rates.

Even in market economies, markets for agricultural labour are difficult to organize on both supply and demand sides. Informal networks with very varied efficiencies in communication are most common. But various organizations concentrate demands in 'feeing' markets for one year contracts, or demands for seasonal labour, migrant labour, or 'settlers'.

16.3.2 Supplies

The numbers who might seek farm work are largely determined by the same demographic and other factors that determine potential numbers of farm managers. Many are themselves potential managers, or are in their households. Generally, it is useful to consider the numbers in agriculture as *remainders* from total rural numbers after deducting those leaving. The 'pull' of non-farm opportunities and the 'push' resulting from farm-economic decisions about numbers have been of great socioeconomic importance especially in relation to hired workers. The determinants of 'pulls' are the economic decisions of non-farm employers as affecting demands for non-farm labour, and the personal decisions of farm employees or potential employees. Much has depended therefore on industrial and other economic growth; types and extents of education and training; urban as against rural living conditions and costs; relative income levels; judgements of future security of income; flows of information; age groups; value orientations of farm workers and potential workers. The determinants of 'pushes' are best understood as the decisions of farm firms, based on farm-economic planning (Part IV).

In many market economies, an *inward* flow of workers to agriculture occurs, particularly during periods when urban unemployment is high. In the USA, this inflow has been much smaller than the outflow, except in the years 1930–32, 1945, and 1946.

Rural areas in many countries (e.g. India and China) have such large numbers of potential farm workers that even at low wages they cannot all be employed. And much of the 'pull' to towns is based on poor information. The prospects of further population increases aggravate socioeconomic problems. Attempts become necessary to provide more productive employment on farms as well as in non-farm activities.

In all countries, the supplies of farm labour should be assessed for 'qualities'. With changing technologies and intensification of production systems, this is especially important. The actual 'prices' paid do, in many areas, vary with the 'qualities', premia being paid for special skills, experience, supervisory duties, long hours, overtime, special efforts at times of peak requirements, and so on. But the practical difficulties of precisely matching demands and supplies of 'qualities' at the right 'prices' are substantial (Table 16.1 and related text).

16.3.3 Prices (rewards) for hired labour

Because numbers and potential numbers are high relative to demands, the rewards for farm labour are relatively low. This assists the outflows to non-farm work. These commonly depend at least partly on this relative lowness, and in very few areas have they fully corrected it. Measures of disparity should, however, allow for non-material values and uncertainties. They have therefore to be interpreted in the light of the value orientations of particular social groups and individuals.

If societies aim to reduce the apparent inequalities, they seem likely to make most progress by a combination of (a) population and health controls; (b) development of non-farm activities and employment; (c) farm-economic planning, improved farm equipment, and education and training of farm managers and labour. In 'market' economies, improvements in the organization and procedures in markets for labour can be made, but will probably have less effect. In almost all economies, the scope for government and social security policies is inevitably limited by costs.

FURTHER READING

*Bishop, C. E. (ed.) (1967). *Farm Labour in the United States*, Columbia University Press, New York and London.

Gasson, R. (1973). Industry and migration of farm workers. In *Oxford Agrarian Studies*, Vol. 2, Agricultural Economics Institute, Oxford, pp. 141–160.

Giles, A. K., and Stansfield, J. M. (1980). *The Farmer as Manager*, Allen and Unwin, Hemel Hempstead.

*Macmillan, J. A., and Gislason, G. S. (1980). Canadian farm operator mobility analysis. *Canadian Journal of Agricultural Economics*, **28**, 11–25.

234

†Mikitenko, I. A. (1978). Inter-relations and development of Labour productivity and remuneration [In Russian]. In *Visnik Sil's'kohgospodarskoyi Nauky*, **5,** 102–106, Ukraine Research Institute of Agricultural Economics, Kiev.
*Tang, A. M., and Stone, B. (1980). *Food production in the People's Republic of China*, Research Report No. 15, International Food Policy Research Institute, Washington, DC.

QUESTIONS AND EXERCISES

1. In selecting a farm manager what 10 skills or attitudes would you consider most important? Explain briefly the biological, work, farm-economic, and socieconomic reasons.
2. In selecting hired workers what 10 skills and attitudes would you consider most important? Explain why.
3. During the last 10 years, in the area you know best, what have been the trends in real terms in (a) farmers' incomes *or* (b) hired workers' wages? What changes in demands and supplies have been the main determinants of these trends?

Chapter 17

Factor markets III: Machinery, fertilizers, seeds, and other factors

17.1 INTRODUCTION

17.1.1 Importance

Why 'biological' and 'equipment' inputs of many kinds are important in securing increased production economically is made clear in Parts II, III, and IV. Illustrations are given for farms of our example areas in Tables 3.4, 9.5, and 10.1. Even in poor countries the inputs bought by farmers are of key importance (e.g. hand tools, some seed, irrigation water, etc.).

We should note that the effects on outputs of some inputs may be much greater than their total values suggest, because only small amounts are needed (e.g. trace elements for soil fertility).

17.1.2 Need for research and development and spread of their results

Parts II, III, and IV emphasize how important to economic production are the 'kinds', 'qualities', and timings of inputs. Chapter 20 considers research, development, and education as affecting these, but here we note that the need for them affects markets for inputs. 'Designs' are important knowledge, and should be supplied along with inputs.

17.1.3 'Lumpy' factors

'Lumpy' factors raise major problems on both the demand and supply sides of input markets, and even for governments. In some farm economic systems (e.g. in the Punjab) a 35 h.p. tractor is very 'lumpy' for a medium-sized farm. An extreme example is the establishment of a major irrigation scheme.

17.1.4 Maintenance and repairs

Another characteristic of many factors is that they require regular, skilled maintenance and repairs. Some of this service flow is provided on farms by farm labour, but many skilled services and spare parts have to be provided in the markets for inputs (e.g. by tractor retailers (dealers)).

17.1.5 Services

Another aspect of development is the increasing use by farm firms of other purchased services. Contract services for mechanized field operations may be purchased. Services may be bought in processing feedingstuffs (e.g. by mobile mill-and-mix contractors). In the early stages of produce marketing, transport, packaging, and cold storage services may be purchased. Also professional services from legal advisers, accountants, and veterinary practitioners are required. Markets for all these services have their own peculiarities.

17.2 DEMANDS

17.2.1 Times and places

All the purchased inputs should of course be available to farmers at the right times and places. This is obvious for repairs, to, say, combine harvesters. But, especially where farm firms are poor and unable to store supplies long without losses, the values of seeds, fertilizers, spray materials, and the like are much lower if they are not provided at the right times and places. Thus in input markets the *time* and *place* aspects of demands as well as the *form* aspects are of vital importance. One method of reducing time and place problems is to establish co-operative societies and buying groups that can bring together the comparatively small and scattered demands. Farmers' demands in particular areas have to be sufficiently similar, and the buying and distributing functions efficiently completed. In the USA farmers' co-operative purchase about 21 per cent of the total feed that farmers buy. The comparable UK figure is about 16 per cent.

17.2.2 Farm-economic sources of demands

The demands of individual farm firms are of course the results of their own farm-economic decisions. Demands can therefore be greatly affected by research and extension education if these are relevant to local biological, work, and farm-economic conditions (Chapter 20). Advertising and salesmanship can help to make the results more effective, but can mislead if ill used.

For some inputs, government regulations help to provide guidance about invisible 'qualities'. Certain measures must be stated; or sales of certain types prohibited. Thus the NPK contents of 'artificial' fertilizers and various chemical analyses of feedingstuffs must be stated in western Europe and North America. Regulations about the description and control of seeds are increasing. Sales of some chemical compounds are banned (e.g. certain poisonous pesticides). Some drugs may be sold only through qualified veterinary practitioners (e.g. many antibiotics in the UK). Strict quarantine tests of imported plant materials and animals are required. Such guidance is increasingly desirable as agriculture develops.

17.2.3   Organization to use 'lumpy' factors

The demands for 'lumpy' factors depend partly on the sizes of farm firms. But the distribution of farms according to size has itself been altered by changes in the sizes and designs of some factors because these have changed the competition for land and land use. (For example, the development of tractors and combine harvesters has in recent decades substantially increased sizes of grain-growing farms in the central Illinois and Saskatchewan areas). Many farmers fear structural changes from mechanization of field work in India and other densely populated countries with small farms (Sections 15.2.1 and 15.2.5).

But it is useful to think of the services of a 'lumpy' factor as a flow that, season by season, is divisible and saleable. Not all the input units of this flow need to be used on the farm of the factor-owner himself. The owner can be (i) one farmer who sells some input units (e.g. hours of work by tractor and driver) to neighbours; (ii) a partnership of two or more farmers; (iii) a larger group, or co-operative society; (iv) a central unit servicing tenants in a large estate; (v) a government agency (e.g. operating an irrigation scheme); (vi) a contractor serving many farmers but without land use himself.

Thus the demands for the 'lumpy' factor may come from various types of organization. Their base is still the decisions of individual farm firms, but these have more scope, within the limits of economic management, because the flow of input units from the 'lumpy' factor is divided up.

Another important part of the demand for 'lumpy' equipment is that from smaller farm firms who buy second-hand machines. Thus, in the UK, there is an economic flow of second-hand equipment from the large lowland farms of the east to the smaller farms of the west.

17.2.4   Loans affect demands

The availability of loans of suitable types affects demands for inputs (Section 18.3). Thus in poor areas where incomes are low and the propensities of farm families to consume their own grain and other produce are high (i.e. income elasticities of demand are high), sufficient funds are commonly not available for purchase of improved cultivar seeds, fertilizers, and spray materials. Replacement of a cow or purchase of even the simplest equipment may not be possible without borrowing. Where these are the general circumstances, the production responses to inputs that are economic are especially valuable from a welfare standpoint. But loans as well as inputs must be available. In richer areas, some purchases are so large (e.g. combine harvesters, buildings) that again loans are required.

17.2.5 Fiscal measures affect demands

In attempting to secure development many governments have reduced by subsidies the costs to farm firms of inputs such as NPK fertilizers and lime, and the costs of some operations such as drainage. Many have subsidized tractors and other equipment or allowed farm firms to deduct abnormally high depreciation on these when calculating taxable incomes. Import duties have been kept low. These fiscal measures obviously affect demand for inputs through the net costs in farm plans and budgets. But also, for some years after introduction, they can have an educative effect because farmers can see the results (e.g. of liming, drainage, phosphate fertilizers). After the measures are withdrawn, this will continue to influence farm-economic decisions. Various government policy questions arise of course. (For example: Could the educative effect be secured at lower cost? Do the fiscal measures favour substitution of capital for labour? Large farms as against small?)

17.2.6 Variations over time

The fluctuations in purchases of replacement and new equipment in developed market economies are related, through farm firms' decisions, to variations in farm incomes. These are determined both by general price level changes and by weather and other causes of variation in agricultural production and the real prices of agricultural produce. Farm firms may also reduce purchases of phosphorus and potassium fertilizers in difficult periods. But to reduce other inputs would commonly not be economic.

17.3 SUPPLIES

17.3.1 Problems in supplying

The above review indicates that these problems are essentially of six kinds. These are concerned with (i) location, sizes of lots sold to farm firms, and timing; (ii) research and development; (iii) communication of knowledge; (iv) loans; (v) costs and pricing; and (vi) structure, conduct, and performance of the whole supply organization. The six kinds are closely interrelated (e.g. the research and development in the design of a new tractor for Indian agriculture (ii) raises questions about sizes of unit (i); advisory work on farm plans and organization of contract work (iii); credit (iv); costs (v); and sizes and policies of manufacturing plants and distributing agencies (vi). We can, however, usefully list the kinds of problem separately, because sound development requires each to be satisfactorily solved.

For different types of goods or service the relative importance of the six kinds of problem and the many details are different. And between and within countries they differ, because the socioeconomic conditions affecting demands and supplies differ. A brief consideration of a selection of input markets serves to illustrate these points.

17.3.2 Selected input markets

Tractors In areas using many tractors, local dealers sell them and provide maintenance and expert repair services. Manufacturers commonly sell only to selected dealers whom they franchise. Sales are promoted by careful designs and much advertising. Dealers assist in arranging loans. Prices are largely determined by manufacturers, and dealers' 'margins' are limited within areas by competition between dealers. This may be judged inadequate in some localities. In areas with few tractors (e.g. in developing countries) distribution problems are great because costs are high, competition limited, expert knowledge and repair skills inadequate, and design not closely suited to local conditions. The training, number, and location of dealers and repairers is therefore of basic importance.

In manufacturing, the economies of scale are so great that the five largest firms make 90 per cent of the tractors in the USA. In the UK, the comparable figure is 95 per cent. The competition between firms is therefore oligopolistic, with much differentiation between tractors so as to segregate demands by special design features, advertising, other promotions of sales, related implement designs and services, franchise of dealers, and so on. This type of competition has been criticised as leading to an excessive number of dealers, manufacturing firms that are too large, unused production capacities, high barriers to new entrants, and high ex-factory prices (Moore and Walsh, 1966). But it can also be claimed that this type of competition has fostered great advances in tractor design.

Where farms are small, progress may be through smaller, simpler designs (e.g. Japanese two-wheeled tractors), and through organization to use tractors as 'lumpy' inputs (Section 17.2.3.)

Research and development are so important and expensive that some governments provide testing and reporting services on new designs. Some also finance research and development work to secure implements for conditions that require special studies but where the eventual total value of implement sales would be too low to induce manufacturing firms to finance such studies.

Repairs of machinery These are now major inputs in mechanized areas (see Table 10.2). In any one locality, those for particular tractors or other complex machines can commonly be obtained from only one source or a very small number of sources. Prices tend to be high because possible new suppliers face entry barriers due to manufacturers' controls of spare parts, and high establishment costs relative to potential local demands.

In many developing countries, some problems are acute over complex machines. Simple repair skills are in surprisingly good supply in some towns (e.g. in India or Nigeria), but scarce in many country localities.

Fertilizers Governments commonly undertake much research, development, and extension work. The larger manufacturers also make experiments and inform farmers and distributing dealers. Investments in research and development to reduce manufacturing costs and secure more concentrated and suitable compounds are also substantial.

The economies of scale are large in the fixation of atmospheric nitrogen, and the making of phosphoric acid and concentrated compound fertilizers. Thus in the UK basic production is dominated by only three firms. Economic problems in locating plants are important because the total tonnages to be transported are large. The opportunities for product differentiation are more limited than in the tractor market, and competition from imported fertilizers may be keen.

Distribution has to be mainly through dealers and amongst these competition is also keen.

In countries such as India, procuring supplies from the cheapest sources within the constraints due to limited foreign exchange presents special problems. Distribution problems concerning location and the small sizes of lots bought by farmers, seasonal timing, knowledge, and credit are together even greater.

Agrochemicals Within the great variety of chemicals for control of biotic factors, many are well established in modern agriculture and are made and distributed under conditions sufficiently competitive and efficient. But, as noted in Chapters 4 and 5, research and development secures many new compounds and these should be carefully used (e.g. new fungicides for particular fungus species with many and variable strains that attack cereals). The costs of research and development are very high and economies of scale in manufacture are substantial. Governments commonly therefore allow producing firms exclusive patent rights for a period of years so that they can sell at prices high enough to repay research and development as well as other costs. Firms therefore differentiate their products through 'design', and by advertising and other promotions. The competition is monopolistic but often fierce. Some government agencies test new products and provide independent advice about them through extension services. Some basic research and development may also be by government-financed institutes. The great increase during recent decades in the use of agrochemicals has been based on this system. Without it, modern agriculture would very probably have been much less productive. But some criticisms of high costs and insufficiently understood biological results can be made. One cost safeguard is that distribution of individual compounds even during periods covered by the patents is generally too costly unless it is done by dealers selling competing compounds.

Seeds Breeding and selecting out improved cultivars is a skilled, lengthy, and costly process (Section 4.4.3). Some governments, including those in the European Economic Community, therefore have a policy on new cultivars similar to that on new agrochemicals – official testing, and, if registered, the granting of special rights for a period of years to secure high revenue for breeders. This ensures a continuous flow of new cultivars of major species for intensively farmed areas in developed countries. For poorer countries, the network of international research institutes and government agencies secures an important flow which is an essential part of the 'green revolution'.

But costs are generally too high for the gearing of genetic programmes to most particular areas, and for adequate testing before sales to farmers. Substantial uncertainties have therefore still to be borne by farm firms (Section 12.4.2).

Other important 'qualities' in seeds are freedom from diseases, pests, weed seeds, and impurities. Vitality and uniformity in germination should also be tested for.

Securing 'qualities' in seeds entails special storage and inventory problems and can provide opportunities for monopolistic competition. Many governments, and firms concerned with processing and marketing (e.g. sugar beet factories, cotton ginneries), find that special services in seed supply greatly increase production and are essential to their own efficiency.

Veterinary services These require research, and skilled professional services in detailed provision. But some provision can be by farmers themselves using purchased chemical and other supplies. The quality of professional staffs, and to a large extent the numbers qualified, is controlled by professional associations and governments.

Special problems arise about how to pay for preventive plans and services to secure full health. Many plans must be for whole countries (even regions). Much preventive work is by government agencies. When plans have to relate to individual farms, they commonly require more understanding and foresight than farm firms yet have. Forward contracts for preventive services from veterinary practitioners are still rare.

Compound feedingstuffs One major problem is to know which 'mixes' will provide, at least cost, the energy and nutrient contents required to supplement those in the roughages and grain produced on individual farms so as to suit their animal production plans (Section 5.2). This requires much analytical work, prompt adjustments of raw material purchases, and many different mixes.

Other problems are about sizes and locations of plant as affected by economies of scale in manufacture, transport costs, and the advantages of local knowledge of demands. In some areas in the UK and the USA, mobile mill-and-mix services that add suitable protein and mineral supplements to farms' own grain now compete with both large and medium-sized compounding mills.

In western Euorpe and the north-eastern USA the distribution of feedingstuffs is commonly competitive. But some differentiation to secure advantages in monopolistic competition is misleading and costly. Vertical integration of compound feed distribution functions with those of manufacture is increasing in the USA. Also increased use is made of contracts to integrate with egg and broiler production.

Petroleum products Because of high effective demands, limited world reserves, and the location of extraction, the Organization of Petroleum Exporting

Countries (OPEC) has, since 1973, been able greatly to raise world price levels. But major undertakings have been initiated in the areas of research and development – to disclose future reserves, to secure greater efficiencies in use, and to find substitutes.

Distribution networks appear to determine margins on the basis of oligopolistic behaviour or, in remote areas, of small local monopolies.

REFERENCE

Moore, J. R., and Walsh, R. G. (eds) (1966). *Market Structure of the Agricultural Industries*, Iowa State University, Ames, Iowa.

FURTHER READING

Abbott, J. C., and Makeham, J. P. (1979). *Agricultural Economics and Marketing in the Tropics*, Longman, London.

Barker, J. W. (1981). *Agricultural Marketing*, Oxford University Press, Oxford.

*Cowling, K., Metcalf, D., and Rayner, A. (1970). *Resource Structure of Agriculture: An Economic Analysis*, Pergamon Press, Oxford.

Rawlins, N. O. (1980). *Introduction to Agribusiness*, Prentice Hall, Englewood Cliffs, NJ.

Roy, E. (1980). *Exploring Agribusiness*, 3rd edn, Interstate Printers and Publishers, Danville, Ill.

Rhodes, V. J. (1978). *The Agricultural Marketing System*, Grid Publishing, Columbus, Ohio.

QUESTIONS AND EXERCISES

1. In what ways are farmers' demands for the factors considered in Chapter 17 increased?
2. In what ways do supply and pricing arrangements tend to reduce purchases of these factors by farmers?

Chapter 18

Factor markets IV: Loans

18.1 INTRODUCTION

The main purpose of loans is to enable firms to produce more. With loans (liabilities) they can secure the use of more assets then they could with their net worths alone. Their 'input mixes' can be better balanced. Their productive lives may start earlier.

Assets can of course be financed in other ways. Established firms may save from their own net flows from 'production' (Figure 10.1). A very large proportion of the assets in agriculture has been financed in this way, and then inherited. Corporations have assets financed by shareholders (e.g. rubber, tea, or sugar plantation companies; private corporations set up by families in areas of western Europe and North America with rapidly changing agriculture and high taxation). Land and buildings may be rented. Equipment use can be bought from contractors. Governments may make grants. But, overall, these ways are inadequate in relation to the total assets that are judged to be economic. The loans used in our example areas are summarized in Table 10.1.

Another purpose of loans is to permit better timing of consumption. Families with incomes that are very seasonal, or subject to big year to year variations, can borrow when they are low and pay back later. Those whose future annual incomes are expected to be high enough can borrow for durable consumers' goods (e.g. washing machines), and pay back later.

Thus loans can be highly important in socioeconomic systems because they affect (i) production and markets for factors and products; (ii) consumption and the reduction of difficulties due to fluctuating incomes. They also affect (iii) savings, because production is increased and some inducements to save are provided.

Governments are commonly concerned also because (a) loans are difficult and costly to provide to farm firms, especially small firms with low incomes; (b) mobilization of savings is related to lending; (c) training and experience in the use of loans can have educative effects that are valuable in national life; (d) social conflicts can arise between borrowers and lenders (between German peasants and Jewish lenders during the 1920s and 1930s; between Indonesian peasants and Chinese lenders in the 1930s and 1940s).

18.2   PURPOSES, FLOWS, RATIOS, AND REPAYMENTS

18.2.1   Purposes and flows

Important aspects of the provision and use of loans to farm firms can be understood from Figure 10.1. At the beginning of the accounting year, an established firm's assets are financed by *liabilities*, and net worth. During the accounting year, new *liabilities* may make some purchases possible. Assets and purchases along with the farm firm's own labour and management provide the 'input mix' that determines the production. What purchases contribute to the 'input mix' depends partly on what proportions of them are directly withdrawn for consumption or flow into end-of-year assets. (For example, fertilizer or equipment delivered late contribute little to the year's production, but more to end-of-year assets.)

Liabilities at the beginning *plus* new liabilities *plus* interest due, result in liabilities at the end of the year, except that some assets or production may be used to reduce this total. But assets are required for production and production has also (i) to pay for many purchases; (ii) to offset depreciation of machinery and other 'lumpy' assets; (iii) to contribute the returns withdrawn by the firm for living; (iv) to pay taxes; (v) to provide savings so that net worth increases. Much therefore depends on how large production is.

Production loans that are successful improve the 'input mix' in quantum and balance so that production is increased to make possible (i) the payment of interest and the repayment of loans that are due; (ii) the avoidance of undesired increases in end-of-year liabilities; (iii) some increases in net worth; (iv) additional returns for withdrawal, to be sufficient inducement to the farm firms.

Loans to permit investment in 'fixed' assets that will provide a flow of inputs over a period of years (e.g. a drainage scheme) will of course be expected to add more to end-of-year assets than to production.

Loans made to sustain consumption are successful if they add sufficiently to withdrawals during periods when these would otherwise be unsatisfactory. The management and labour of the firms are sustained, so that later production is adequate to repay the loans with interest.

Loans should be closely related to knowledge of production and market conditions so that farm plans are well made and executed. Due regard should be paid to 'quality' of inputs and to proper timing. In other words, *what* loans are for, *how big* they are, and *when* they are made and used are all important for their success. And what it is planned to produce should be marketable.

18.2.2   Ratios

Loans differ from other purchased factors (to which they may be related) because (i) loans may not be repaid if production is inadequate for any reason (lenders are therefore commonly much concerned about uses); (ii) loans usually provide opportunities to make greater mistakes in farm plans and their execution than do purchases that are limited directly by net worths alone; (iii) loans can also lead to rates of withdrawal for consumption that cannot be sustained.

The variability of production and prices from year to year is commonly so great for individual farm firms that it adds substantially to the probabilities of short term production loans being unsuccessful. And long term loans may be unsuccessful if product prices fall relative to the prices paid for assets such as land, and irrigation schemes, or the expected productivity of such assets is not achieved.

Thus wise borrowers try to ensure that:

(a)   Their farm-economic plans are well chosen, with due regard to risks and uncertainties as well as to average 'profits'.

(b)   In the execution of plans, risks and uncertainties are reduced so far as is feasible and economic.

(c)   Despite the remaining risks and uncertainties, loans can be repaid when due, and interest payments completed, because the relationships are sound between (i) liabilities and interest payments and (ii) net worth and production *less* (purchased inputs, depreciation, required withdrawals for consumption, and taxation).

Lenders are also concerned with these three aims because they all affect the *repayment capacity* of borrowers.

In general in world agriculture, the main ways of maintaining repayment capacity have been by (a) and (b), by borrowing little to purchase little, and by varying withdrawals according to variations in production. Small family farms have been the structural foundations for this strategy. Thus even in the 1970s in India, relatively few farmers had borrowed enough to have to pay interest equivalent to more than 1 per cent of their total annual expenses (World Bank, 1975). Even in our example Aberdeen area, 25 per cent of the farmers who were tenants in 1978–9 had liabilities that were less than 10 per cent of their assets. But other areas require much higher loans, particularly to finance ownership of land and buildings and farm firms that are 'expanding'.

18.2.3   Repayment capacity

Some farm firms seem unable to avoid relatively high liabilities. Their plans, or executions of plans, are inadequate, or their misfortunes exceptional. Some may even prefer balance sheet ratios that show continuously high liabilities and indicate too little ability during difficult periods to use reserves and save on withdrawals. So in 1970 in India, 7 per cent of the loans due to be repaid to Primary Co-operative Credit Societies were not repaid, and 34 per cent of the total loans outstanding were in arrears. And in 1978–9 in the Aberdeen area 17 per cent of the tenants had liabilities equal to more than 45 per cent of their assets. Table 18.1 contrasts a firm that used loans well with a firm that was comparatively unsuccessful.

Table 18.1 Contrast between two farm firms in net worths and use of loans:
Aberdeen area, 1968–78.

Farms and Periods[a]	Farm area	Net worth	Loans	Total assets[b]	Management and investment income	
					Per hectare	Per £100 total assets[b]
	ha		£ per hectare[a]		£	£
Farm A						
1968–70	69	91	86	177	17	10
1972–74	93	160	24	184	30	16
1976–78	87	319	45	364	26	7
Farm B						
1968–72	28	109	32	141	−29	−21
1972–74	29	91	9	100	−24	−25
1976–78	29	91	7	98	7	7

[a]Periods of three harvest years. The averages per year are set out. The money figures are all in pounds sterling of 1968–70 value.
[b]Excluding land and buildings.

Repayment capacity can be greatly reduced by ill-health, accidents, and malnutrition (Section 7.5). These aggravate difficulties for both lenders and borrowers, particularly in poor countries with inadequate health services and insurance. They are among the reasons for older social arrangements within extended families and smaller communities.

18.2.4 Willingness to repay

Because repayment of loans depends on willingness to repay, lenders assess the personal characters of the borrowers as well as their abilities and energies as producers. This willingness may be reduced by (i) their propensity to consume; (ii) their preferences for continuation of liabilities as against the alternatives of working harder to produce more, reducing assets, or saving more and reducing consumption; (iii) (in some less developed countries) their lack of understanding of the purposes and conditions of loans.

Decisions in favour of continuing liabilities are now common in many types of economy because the total assets required to make use of what are judged to be the best technologies are much greater than in the past.

Lack of understanding of loans is common in Africa, and in Latin America. In some areas, they may even be regarded mainly as tokens of the social status of the lenders.

18.2.5 Security for lenders

As a safeguard against non-repayment, lenders may require collateral rights in the borrower's assets. These are commonly mortgages on land and buildings. Lenders to tenants may require 'security interests' in other assets (e.g. chattel mortgages on movable property). Real estate mortgages are easier to use as collateral and usually less costly, because the loans are for larger amounts and longer periods, and the assets are not movable. But where land reform and size-of-farm policies aim to create and maintain many small farms owned by their occupiers, and where there are tenants, other 'security interests' may have to be used as collateral (e.g. liens on sales by borrowers through co-operative societies).

18.3 DEMAND SIDE

Borrowers have some requirements that are specially noteworthy:

(1) The conditions of loans and rates of interest and other costs should be so stated that loans from alternative sources can be validly compared.

(2) The timing of repayments of loans for production should be related to the flow of net products from the assets they finance, and before these assets become obsolescent. Lenders should be permitted to make repayments earlier if they wish, and interest charges should be on only the net loans outstanding.

(3) Because net income flows are so variable from year to year, lenders should avoid damage to 'input mixes' by occasionally agreeing to delays in repayments or to special new loans.

(4) Lenders should understand all the loan needs of borrowers so that farm economic plans can be complete and household consumption plans consistent with them.

Borrowers of loans for consumption should know particularly: (a) what 'security interests' the lenders have (e.g. they may still be owners of some durable consumer goods that have legally not been sold but lent; they may require produce to be marketed through them); (b) how long the flow of satisfactions will continue from the goods bought, as compared with the period for repayments; (c) when net income will probably be sufficient to repay the loan with interest.

Because of all the biological, work, and farm-economic variables that we have surveyed, the loans that are appropriate for individual firms and households tend to be unique to them. But certain loans are of special concern to governments: for young men to start their farm firms; for purchase of farms for amalgamation from old men who would retire; for other types of farm improvement; for machinery to be owned by groups of farmers; for co-operative societies or other groups to improve their bargaining power; for other marketing improvements; for part-time farmers; for low-income farmers, e.g. those not experienced in commerical transactions; for tenants in settlement schemes; for new owners who, before land reforms, had loans from landowners. *Supervised credit schemes* were introduced for low income families in the USA during the 1930s and are now used in developing countries, particularly in Latin America. They bring together loans and technical, farm-economic, and household management advice to ensure that loans are used well.

Despite the length of this list we should note that, in many industrialized economies, loan needs and problems have received less attention from governments and publics than have interventions in product markets and subsidies on products and other factors. And in 'developing' economies, the World Bank estimates that 70–80 per cent of small firms still have no access to modern lending institutions.

18.4 SUPPLY SIDE

18.4.1 Funds available

The funds available for lending to agriculture and the interest rates required are largely determined in the general money markets of individual countries. The savings available for lending depend on levels and distributions of income, propensities to save, and mobilizations of savings. Most of those lending to agriculture consider the net returns that could be secured from alternative borrowers. Many lenders have responsibilities to those who have deposited savings with them, and these responsibilities must be met if savings are to be made available. Governments may increase the supply of loans for agriculture by subsidies and guarantees of repayments. The reasons for such government actions are essentially judgements that the supply is inadequate in relation to the goals of agricultural policies.

18.4.2 Costs of provision

The costs of providing loans are determined by (i) basic interest rates in money markets; (ii) costs of 'delivery'; (iii) costs of monitoring the use and productivity of loans, and of actions when interests or repayments are overdue; (iv) costs of loans never repaid. In 'developing market economies' all these tend to be high. Thus the World Bank has stated that a basic interest rate in real terms of 15 per cent becomes more than 21 per cent to farm borrowers even when the lenders are exceptionally efficient, investing their capital during the whole year, holding administrative costs to 3 per cent, and defaults to only 3 per cent. Many lenders cannot be as efficient. And in some localities the competition between lenders is imperfect, so they charge still higher rates.

18.4.3 Organization

The organization of the supply side of markets for loans has tended to evolve in most countries, to suit their particular socioeconomic conditions. The two major groups of function concern (a) mobilization of savings and government funds; (b) delivery to borrowers, and related collections of interest and repayments.

A typical institutional organization is (i) a central bank; (ii) a special bank with responsibilities for finance of agriculture; (iii) area branches of the special bank, or co-operative societies, responsible for actual deliveries and collections. Other deliveries are by smaller, commercial firms, such as (i) merchants supplying fertilizers, equipment, and other factors; (ii) finance companies arranging hire purchases; (iii) banks and other institutional lenders; (iv) small moneylenders; (v) relatives or other individuals.

In the USA institutional lenders include commercial banks; life insurance corporations; the farm credit system (including the banks for co-operatives); the Farmers' Home Administration; the Commodity Credit Corporation; the rural Electrification Administration; mutual savings banks; and finance corporations providing loans for purchase of equipment and animals.

In the United Kingdom, commercial banks and their branches provide most of the loans, and the government-sponsored special institutions and co-operatives provide comparatively little.

In Germany, small credit associations were developed after 1854 by Raiffeisen and others, and they became world famous. Their knowledge of production and marketing conditions and of individuals and social relations was intimate in their own localities. They could at low cost judge which loans would be sound and avoid the need for detailed collateral security. They secured central funds on the basis of aggregated local demands and pooled risks and uncertainties. They encouraged thrift and saving, and mobilized local savings.

The difficulties and high costs of delivery systems particularly in developing countries have led many governments to try to develop adequate networks of effective co-operatives. Following land reforms, multi-purpose societies are favoured – able to sell factors, market products, make loans against liens on product sales, and assist with supervised credit schemes. In practice serious difficulties retard progress where (i) understandings of loans and their relation to farm-economic plans and management are inadequate; (ii) well trained and reliable staff are unavailable, or too expensive; (iii) record keeping and management of the multi-purpose business are too complex; (iv) members of the co-operative take too narrow a view of its management problems; (v) governments, in attempting to overcome these difficulties, impose too many bureaucratic controls, inject too many subsidies, lead to many decisions becoming 'party-political' or corrupt, and so 'give the kiss of death'. (See also Chapter 22.)

Area branches of agricultural development banks or similar government agencies are liable to suffer like difficulties.

These are reasons why 'informal' delivery systems are still so important as compared to 'institutional' ones.

18.4.4 Changes with intensification

The intensification of farm production systems has in many countries made reliable judgements of farm plans and efficiency more difficult to achieve.

Modern institutional lenders rightly require, when considering new loans, substantial evidence from a farm firm about (i) balance sheet and net worth changes; (ii) operating statements for recent years; (iii) an overall budget for a year ahead; (iv) more detailed partial budgets related to the purposes of the proposed production loans; (v) a cash flow projection indicating receipts and expenses, risks and uncertainties, and opportunities to pay interest and repay the loan.

On such economic information depends the avoidance of arrears of interest and repayments and the securing of more total funds for lending. But where farms are small, and individual loans are small and for only short periods, the costs are excessive. Thus lending to large and lending to small farms should be separated in many economies. The problems in designing fully satisfactory loan delivery systems for small farms in poor countries have still to be solved.

In industrialized countries the major problems to be solved are (1) how to finance new firms to take over from old when asset values are increased by modern technological requirements, larger farm sizes, conservation needs, and inflation; (2) how to avoid the heavy borrowing and high interest charges that threaten too many firms during and after periods of low incomes.

REFERENCE

World Bank (1975). *Agricultural Credit*, Rural Development Series, World Bank, Washington, DC.

FURTHER READING

*American Institute of Banking (1969). *Agricultural Finance*, American Bankers Association, New York.

Barclay's Bank (1980). *Finance for Farmers and Growers 1980–81*, Barclay's Bank, London.

†Brake, J. R., and Melichar, E. (1977). Agricultural finance and capital markets. In *Survey of Agricultural Economics Literature* (L. R. Martin, ed.), North Central Publishing (for the American Agricultural Economics Assocation), St Paul, Minn., pp. 412–494.

*Desai, S. S. M. (1979). *Rural Banking in India*, Himalaya Publishing House, Bombay.

Hale, C., Herndier, G., and Rana, M. (1979). *Farm Credit Management*, Marketing and Economics Branch, Saskatchewan Agriculture, Saskatoon.

*Jodha, N. S. (1978). *Role of Credit in Farmers' Adjustment against Risk in Arid and Semi-arid Tropical Areas of India*, International Crop Research Institute for the Semi-arid Tropics, Patancheru, AP, India.

Lee, W. F., Nelson, A. G., and Murray, W. G. (1980). *Agricultural Finance*, Iowa State University Press, Ames, Iowa.

*Penson, J. B., and Lins, D. A. (1980). *Agricultural Finance: An Introduction to Micro and Macro Concepts*, Prentice Hall, Engelwood Cliffs, NJ.

Roy, E. (1980). *Exploring Agribusiness*, 3rd edn, Interstate Printers and Publishers, Danville, Ill.

†Scorbie, G. M., and Franklin, D. L. (1977). The impact of supervised credit on technological change in developing agricuture. *Australian Journal of Agricultural Economics*, **21**, 1–12.

Surridge, B. J., and Webster, F. H. (1978). *Cooperation, Thrift, Credit, Marketing and Supply in Developing Countries*, Plunkett Foundation for Cooperation Studies, Oxford.

QUESTIONS AND EXERCISES

1. For a defined group of farmers (other than any in Table 10.1), set out the average balance sheet for a date during the last 3 years. Indicate how much liabilities vary relative to net worths on individual farms.
2. What are the main risks and uncertainties that borrowers and lenders of agricultural production loans have to bear in mind in the region you know best?
3. List the sources of loans to farmers in a particular region or nation, and indicate their relative importance.
4. What proposals are discussed for improvements in lending to farmers in the nation you know best? What are the reasons against such proposals?

Chapter 19

Infrastructure I: Law, communications, storage, money and public finance, and statistics

19.1 INTRODUCTION

The foundations of socioeconomic systems – their infrastructures – affect agriculture through markets, as we have noted in Chapters 15–19. But other effects can be better understood if we study parts of the infrastructures directly. In this chapter, therefore, we survey briefly some basic components of socioeconomic systems as they affect agriculture – law, communications and storage, money and public finance, and statistics. These are included within the 'boxes' of Figure 13.1 labelled 'infrastructure' and 'central and local government', and they affect those labelled 'capital markets', 'saving and investment', 'taxation', and 'conservation', as well as other 'boxes' that we have already surveyed or will survey in later chapters. The provision of water and power, the protection of human, plant, and animal health, and controls of land use are noted in this chapter under law or public finance or both.

Because they are so fundamental, the components surveyed here concern industrialized economies as well as developing economies.

19.2 LAW

19.2.1 Purposes

In prehistoric societies a basic need was security against outside raiders seeking plunder, slaves, natural resources, and political power. This need can be seen as still basic in tribes that are today largely self-sufficient. Within such tribes, detailed customs govern their internal relationships to reduce thieving and conflicts over food, sex, land, and other matters. The properties of small groups, families, and individuals are protected by customs (mores). In history, such customs were the foundations of written law.

Written law is essentially the common force of society 'organized to act as an obstacle to injustice and to substantiate the natural right of defence of persons, properties and liberties' (Bastiat, 1850).

Even after elaboration in ancient Rome, the purposes of law, and of social organization throughout feudal times in western Europe, were very largely improvements in security, external and internal. But by the fifteenth century, commerce also had become important. Further elaboration of the law began and has continued right up to the present day. The definitions of injustice have been greatly extended and refined. The reasons of most significance for agriculture can be briefly surveyed in five groups (Sections 19.2.2–19.2.6 below).

19.2.2 Revenue collection

Throughout human history, problems have been increasing in deciding how and by how much 'persons, properties and liberties' should be taxed, so as to provide revenue for government. These problems have always raised questions about uses of the revenue; costs of collection; and fairness as between the tax payers whether it was labour, military services, or money that they paid. Increasingly questions have been asked about effects on production, consumption, savings, and investment.

In addition to revenues required to make the law itself effective, greatly increased sums have been used to provide government services, subsidies, and capital for government firms (e.g. for forestry, coal mining). A great and expanding body of law on taxation has been required. Agriculture is everywhere affected, although the details differ between countries concerning such items as poll taxes; land taxes; taxes on factors sold to farm firms; levies on produce processed, exported, or imported; income taxes; capital gains taxes; capital transfer taxes; wealth taxes; and so on.

Governments should try to ensure not only that taxation is appropriate in economic terms but also that it is lawful in that its net results in modern contexts are accepted by societies.

19.2.3 Environments of firms

Security Thieving, bribery, graft, nepotism, and other forms of corruption all affect prices of factors and products, output–input relations, time periods over which investments can be expected to yield returns, the supply and quality of factors, the width of markets, and perhaps most important, the uncertainties of individual firms. In some circumstances, small widely recognized payments to officials to secure their attentions to particular work may have the effects of simple fees, necessary because the officials are underpaid from government revenues. But more commonly, without laws made effective against thieving and corruption, the environments of firms are too insecure for modern, economic production.

Very early examples of laws providing greater security were standardizations of weights and measures. Modern examples are in regulation of (i) accountancy practices, to reduce embezzlements; (ii) government staffing and staff promotions.

Security against human, plant, and animal diseases is increasingly recognized as important and is the aim of many laws.

Facilitating economic activities The standardization of weights and measures is also an example of a wide variety of laws that aim to facilitate buying, selling, and other contracts between firms. Disputes are made subject to arbitration, and rights to compensation are better defined.

Controlling types of firm and imperfect competition Before industrialization in western Europe, firms were small, but because transport was costly many non-farm firms could secure advantages from imperfect competition. Some of the earliest attempts to control firms were therefore local government laws granting monopoly powers to craft guilds with the intention that they would obtain products of assured quality, for sale at negotiated prices.

With more transport, commerce, and industrialization, types of firms had to be more fully defined in *company law*. These types include individual proprietorships, partnerships, joint stock companies (both private, and public in that their shares can be marketed through stock exchanges), and co-operative associations.

Three developments have required further elaborations of great significance for agriculture:

(a)   The high costs of research and development have led to laws on patents (e.g. for agricultural chemicals and farm machinery) (Section 17.1).

(b)   Tendencies to other types of imperfect competition (e.g. through creation of very large firms or 'unfair trading practices') have had to be legally countered.

(c)   Increasing vertical integration by contracts (largely controlled by retailers or processors) has required legal constraints, with related attention to any countervailing arrangements for horizontal integration (e.g. farmers' co-operative associations).

For some selected functions, the economies of large-scale organization, or national security interests, have in many countries appeared to favour the establishment of firms managed by government agencies. Some have complete monopolies; some do not (e.g. broadcasting corporations; forestry commissions; communications, railroad, electricity, and coal mining firms) (see Section 19.2.5 below).

Land tenure Most countries have elaborated laws about land tenure because they wish, with differing and changing priorities, to ensure that the socioeconomic functions of land tenure arrangements (Section 15.2) are carried out.

Labour laws   Many laws relate to the conditions of employment of labour, including those about wage rates, safety and insurance, hiring, housing, training, retirement, trade unions, dismissal, racial and sex discriminations, and migrations.

Banking and credit laws   Laws have been established to protect borrowers, savers and depositors, and lenders in ways that will safeguard their economic interests in fair bargaining. Thus in the USA the establishment of the Farm Credit Administration and related bodies, and in Canada of the Farm Credit Corporation required laws. In some African countries there are strict laws about money lending.

19.2.4   Relations of firms to consumers and communities

Laws to protect consumers against misleading and unfair practices by producers and retailers have a long history. Such laws are recognized as increasingly important (e.g. to regulate the use of hormones in animal production (Section 5.7) and of preservatives and other additives in food processing; also the packaging and labelling of foods).

Other legal protections of communities are against firms whose actions have costs that would, in the absence of the laws, not have to be fully met by the firms (e.g. pollution of water courses; overgrazing of common land; damage to forests, recreational facilities, wildlife, and landscape).

19.2.5   Development of whole economies and reduction of fluctuations and inequalities

In industrialized countries major socioeconomic problems result from (i) fluctuations in economic activity and unemployment and (ii) inequalities of incomes. The poorer countries are affected by the fluctuations, by their desire for development, and by their own inequalities of wealth and income. Increasingly governments have tried to reduce these problems and have made many laws to this end. (See Section 19.4 below.)

Money   Agriculture is greatly affected by shifts in the general price level, unemployment in non-agricultural activities, and related changes in government expenditures, taxation, subsidies, interest rates, and international trade. All of these are interrelated with monetary policies which are also backed by law. (See Section 19.4 below.)

Monopolies and state enterprises   Monopoly powers, local or national, have been granted by law where governments have thought that they would foster development. Government agencies are able to act as firms in development (e.g. in land settlement schemes in Malaya; the Water and Irrigation Department in Pakistan; in exporting cocoa in Nigeria). Some of these agencies are heavily subsidized. Some have been used to obtain government revenues through levies.

In areas where governments seek integrated developments, laws may require co-operation (e.g. in river valley projects for soil erosion and flood control, hydroelectric power, irrigation and drainage). Also in smaller areas, legal powers may be given to majorities of farm firms, requiring minorities to co-operate in soil conservation programmes.

Subsidies Subsidies are now injected into socioeconomic systems in many ways. They all require laws. But because subsidies are not possible without taxation or inflation, governments must ensure that the net overall results are acceptable to society, otherwise they have no sound foundation in law.

Control of locations Laws are used to influence the location of industry and urban developments. They thus affect transfers of land from agriculture, and transport to markets of farm produce.

19.2.6 Church, military, and international laws

Substantial sectors of the law in many countries deal with Church and military affairs and these can affect agriculture (e.g. land tenure, requisitioning for armies).

International laws can also have important effects on security, markets, plant and animal health controls, taxes, subsidies, and other matters.

19.3 COMMUNICATIONS AND STORAGE

19.3.1 Movements of information, ideas, and people

These are essential to the adequate working of socioeconomic systems. Movements are required to make laws well, and to carry them into effect. National ideologies have required transport for their build-up and maintenance (e.g. in the USA, Canada, Nigeria). The making of roads to isolated communities has often started many social and economic changes in them (e.g. in Nigeria). Restrictions on communications between nations that were intended to protect political ideologies significantly retarded technological and economic changes (e.g. in China in the eighteenth and nineteenth century, and in 1962–72).

19.3.2 Transport of products and factors

Movements of goods Movements of goods are essential if the advantages of trade within and between nations are to be secured.

Thus in the forest areas of West Africa, before roads and railways were built, transport was so difficult and costly that the only exports were slaves, gold, ivory, and pepper. Now comparatively bulky products, such as ground-nuts from northern Nigeria and nuts and oil from the 'palm belt' are transported. Table 19.1 gives some examples of the high costs of transport by human labour (porterage) and pack animals and of how much they were lowered by railways, roads, and motor vehicles. Such reductions made economic the 'bringing in' of large areas to commercial production in all regions of the world. This affected both export trading and trading between areas within regions (e.g. between the forest and savannah areas of Nigeria). It also made possible wider distribution of non-agricultural goods (e.g. cement, fuel).

Table 19.1   Costs of transport by different methods: India and Nigeria.

Method	India	Nigeria	India	Nigeria
	year	year	kilograms of grain equivalent, per tonne-kilometre	
Porterage	1937	1926	8.2	6.5
Pack donkeys	1958[a]	1929[b]	6.0[a]	4.7[b]
Carts	1936	...	3.3	...
Railway	1930[b]	1926	0.6[b]	0.4
Motor vehicles	...	1926	...	2.6
on road	1960	1955	0.3–1.8[c]	0.8

Source: Extracted from Clark and Haswell (1967, pp. 184–188).
[a]Pakistan.
[b]China.
[c]Agricultural goods, lowest rate, to industrial goods, highest rate.

Effects on price ratios   Costs of transport affect the production plans of farm firms through price ratios as well as through price levels. The lower the transport costs, the better are the prices in market centres reflected in remote areas. Therefore choices of products and 'input mixes' are better guided by the product: product ratios, factor: factor ratios, and product: factor ratios.

Effects in reducing fluctuations in supply and prices   Reductions in transport costs have other effects that are of great importance in economic development of agriculture. The linking of local markets to wider market networks tends greatly to reduce fluctuations in local prices and to stabilize price ratios. It also helps to reduce uncertainties about food supplies, and supplies of farm inputs. Thus uncertainties in farm planning are reduced. In China when transport was by wheelbarrow or pack animals, localities struggled to be largely self-sufficient in foodstuffs. But year to year variations in weather made some localities seriously short even when, not far away, others had temporary surpluses. Transport is therefore an essential basis both of food security and of economic farm planning.

Network and integration   Costs of transport depend not only on weights and distances, but also on the volume, perishabilities, and difficulties of handling of the various goods; the number of times handled; handling equipment; the nature of the terrain as affecting road or railway building and maintenance costs or the depth and flow of rivers; power costs; capital investments already made; and management.

In general, road, rail, river, pipeline, and air methods each have particular advantages and disadvantages. Both individual firms and governments need to co-operate if the networks are to be impartially co-ordinated to make the overall results economic (e.g. ensuring that heavy and bulky products are usually transported long distances by railways or rivers). There should be neither harmful overlapping nor restrictions.

Benefits and subsidies   The total socioeconomic benefits of transport are judged by many governments to be greater than the sums that users would pay for transport. Subsidies are therefore paid to maintain transport networks. In some circumstances these can be justified on economic grounds. But some transport programmes of governments appear impossible to justify economically (e.g. large prestige highways and airways, rather than many more miles of rural roads; protection of old networks against economic and new methods).

19.3.3   Storage

Infrastructures should permit the transfer of agricultural products and factors over time by storage as well as over space by transport. The large grain stores in the prehistoric Indus Valley and Inca civilizations were necessary not only to facilitate taxation but also to provide security of food supplies. Even in wide modern markets (e.g. the international market in feed grain) large storage capacities are required. Storage arrangements are closely related to those for transport. They also depend on technological developments (e.g. in the control of insects, and other infestations; refrigeration; dehydration; canning; mechanical handling).

Much storage is provided by individual farm firms and by processing and trading firms. But governments also provide much for greater food security, and in attempts to raise and stabilize farm incomes. Both firms and government agencies have economic problems in deciding the locations and sizes of stores, and the labour and equipment to use in them.

19.4   MONEY AND PUBLIC FINANCE

19.4.1   Basic functions of money

Agriculture relies heavily on money in buying and selling, accounting, and planning ahead. We have noted that some recording and planning has to be of physical quantities (e.g. NPK for crops, energy in animal diets). And some values affecting farm-economic plans are difficult to express in terms of money (Section 12.1). Even so, virtually everywhere money is used as the '*medium of exchange*', and the principal '*unit of account*'.

Other basic functions are to provide a '*store of value*' and a '*standard of deferred payments*'. Thus assets deposited in banks as money savings are later repaid as money. Most loans are recorded in money terms, and eventually repaid in money. Agriculture in any particular country is much affected by how well its money performs these functions because agricultural processes require time, and substantial assets, and considerable use is made of loans.

19.4.2 Basic functions of public finance

In addition to law, communications, storage, and money, various other *social wants* may be regarded as basic – e.g. reliable drinking water, sanitation, control of public health, floods, fires, and compulsory schooling. The provisions benefit all, and cannot be well financed and controlled without government.

Some of these basic wants are better satisfied by central and some by local governments, and some of the actual work (e.g. in flood control, and even schooling) may be contracted out to firms. But whatever the division of responsibilities, effectiveness is essential to sound agriculture.

19.4.3 Other functions of money and public finance

In modern socioeconomic systems, governments attempt many other types of function. We noted briefly in Section 19.2 the main types of most significance for agriculture.

Inevitably, wherever the functions of government are expanded, governments have to increase taxation and/or borrow more. Table 19.2 summarizes central and local government budgets in Scotland.

Although such budgets show the costs of some services and subsidies, they do not define other costs (e.g. of international trade controls; agricultural research in universities). Nor do they measure benefits. Much deeper analyses are required, beyond the scope of this book. We should, however, note here again that agriculture is affected directly or indirectly by many government actions, and therefore 'agricultural policy' should not be narrowly defined. General fiscal and monetary policies can often have far greater consequences for agriculture than specific actions limited to agriculture.

19.4.4 Unstable general price levels

Serious consequences for agriculture result from increases in public financial commitments that cannot be met without both high taxation and substantial inflation (i.e. reducing the value of money by more than about 3–5 per cent a year and raising interest rates). Money serves then less well as a store of value and standard of deferred payments. Young men find more difficult the accumulation

Table 19.2   Government expenditures:[a] central and local governments, Scotland
1979–80.

Total current expenditure per head of population	£1335
	per cent
Composition of current expenditure	
Law and order	3.7
Education, research, libraries	17.7
Roads and transport	5.5
Water, sewerage, recreational, and other environmental	7.3
Trade, industry, employment	
Services to labour market	1.5
Other, including job creation	3.1
Agriculture, fisheries	
Subsidies	1.7
Services and inspectorate	1.0
Forestry	0.6
Health service and other personal social security	17.8
Social security (pensions, child benefits, etc.)	27.1
Housing	10.8
Tax collection	1.2
Other public services	1.1
Total	100.0
Capital grants and net lending per head of population	£40

Source: Scottish Office (1981).
[a]Excluding national defence, overseas services, and other expenditures of the UK that cannot be apportioned to different regions.

of the net worths required to start as farm firms. Many older men face more difficulties in making real savings. Thus the structure of farm firms that is suited to a socioeconomic system with lower taxes and little or no inflation may within a decade or two be altered by high taxes and inflation.

Such results are greatest where governments have been led to increase expenditure greatly for the purpose of compensating for 'business cycles' and so sustaining fuller employment but where in fact much unemployment was due essentially to different causes (e.g. needs to improve management skills and energies). Also inflation has serious consequences in developing market economies where governments have not secured increases in tax revenues and loans that together are adequate for their ambitious programmes for investments in infrastructures and industries, and for military and political prestige. In such circumstances, the prices paid to farm firms for their main products may be kept relatively low as a form of taxation, with serious consequences (e.g. in Iran; see Section 23.5).

Falling general price levels in the 1920s and 1930s seriously affected agriculture because the ratios between flexible product prices and less flexible factor prices became unfavourable, and loans became difficult to repay. In recent decades in many countries taxes, interest rates, and the urban wage rates negotiated by trade unions were again high relative to farm product prices. Therefore, even when general price levels rose fast, price and cost *ratios* for farm firms were somewhat similar to those experienced in previous deflation periods (Table 14.2).

19.4.5 Methods of taxation

These are important because they *help* to determine (i) the net revenues to governments; (ii) the criteria on which the decisions of farm firms and market firms are based; (iii) other allocations of resources within economies; (iv) the distribution of tax burdens and therefore of incomes and net worths; (v) some of the effects of budgets on the fluctuations of economies. Poor agrarian countries cannot, at reasonable cost, secure much revenue by income taxes levied on multitudes of small farmers and petty traders, but must use land taxes or export duties or, where cash sales are low, poll taxes (levied simply on numbers of adults). In countries with larger farm firms, income taxes can produce more revenue at lower collection cost, less effect on the use of resources, but some on the distribution of net real incomes as between richer and poorer farm families.

A major aim is commonly to secure high net revenues with least effects on incentives to produce, save, and invest for the future. How this is achieved should be judged by individual governments differently according to the socioeconomic conditions of their countries. Satisfactory results cannot be achieved unless the economic plans of farm firms and marketing firms are comprehended, and their dynamic possibilities well understood. Thus when the inflation rate is substantial in industrial economies, the methods of calculating taxable incomes of farm firms should not make too difficult further economic increases in equipment on farms. Allowances for depreciation or investment allowances should be adequate. In developing countries where mechanization for large-scale unirrigated grain production by big landowners can render some tenants landless, and reduce employment on farms, the various taxation methods should together help to retard such changes in mechanization and related changes in land tenure and employment, so that they accord better with changes in non-agricultural employment. A well designed tax plan may, in some circumstances, prove adequate without sweeping land tenure reforms. In some developing countries (e.g. Nigeria) the simplicity of export duties (e.g. on cocoa) should not obscure the fact that such taxes are not 'progressive'. Poor farmers pay as much per tonne as rich. In all economies, taxes on farm firms should help to reduce rather than increase the high variations from year to year 'profits' after tax (see Sections 12.1 and 12.2). Over-taxation has special dangers where farm-economic decision making for the future is not well understood.

The division of responsibilities between central and local governments should favour appropriate collections of revenues for local purposes, so that they may be better understood and controlled. But where, as in the USA and the UK, land values rise above the actual productivities of the land in agricultural production, care is necessary to avoid over-taxing agriculture by property taxes (local rates).

19.4.6 Subsidies and tax allowances

Many governments pay subsidies on agricultural products and on inputs or operations (e.g. fertilizers or drainage). Some governments subsidize amalgamations of farms and retirements of some farmers. Some even subsidize changes in land use away from tilled crop production.

All governments that tax incomes and property have to define what are the acceptable methods of accounting and valuation. Changing these methods provides different 'allowances', and the results may usefully be considered along with subsidies.

Individual governments have to judge what subsidies and allowances will help to achieve their goals. They must therefore have knowledge and understanding of farm-economic systems, their dynamics and potential. Thus we have noted (Section 17.2.5) that subsidies on factors (e.g. phosphate fertilizers and lime in western Europe) can be used to convince farm firms to use economic amounts. But full success depends on being able to withdraw the subsidies after only a few years because the educative effect has been augmented sufficiently by extension education (Section 20.2) and services (e.g. in soil testing; Sections 4.9.7 and 20.2). Otherwise there are unnecessary government expenditures and misuse of resources by some firms. We have also noted (Section 14.3.1) that product subsidies can be used to reduce year to year variations in farm incomes, but political pressures tend to make objective management of subsidies for this purpose very difficult. Within a few years misdirection of agriculture may result, and uncertainties return (Section 23.5). Other examples of how subsidies and allowances can misguide farm firms are: the continuing long-term subsidization of irrigation water and fertilizers without ensuring maintenance of good soil structures (Sections 4.9.4 and 4.9.6); the heavy subsidization of grain storage, while keeping official prices of grain sold by farmers low and not varying seasonally; heavy subsidization of high yielding varieties and fertilizers without adequate drainage and pest controls.

19.4.7 Finance of conservation

Public finance should help to ensure that the production plans of farm firms conserve water, soil, natural vegetation, and wildlife and other natural resources, and landscape and recreational facilities (Sections 6.5 and 6.6). This means essentially that social rather than individual firm valuations are placed on products and factors throughout the relevant *present value equations* (Section 12.4.1) and that products or factors important socially but 'external' to individual firms are included (Section 19.2.4). Also it means that the discount rate and time periods considered are those of society rather than those of individual firms. Thus, for example, a study of nine farms in the Corn Belt of the

USA showed that if farm planning took no account of soil erosion and did not restrict the use of nitrogen fertilizers, the results were substantially different than when society's wishes to conserve soil, water courses, and the energy used for nitrogen fertilizers were taken into account (Table 19.3). The practical and politically acceptable ways of doing so, and of subsidizing farm firms to take the right decisions, have yet to be found in most areas. Compulsion has been attempted in some countries (e.g. Kenya, Nigeria), but with limited success. In countries with collective farms, conservation plans may be even more difficult to secure, because the effective discount rates in much farm planning may, in practice, be far higher – the time periods far shorter – than societies assume or wish.

Table 19.3   Estimated annual average soil losses, nitrogen uses, and net revenues under three farming policies (nine example farms, Corn Belt, USA; forward plans for 100 years from 1977).

Policies on special controls of soil erosion, and nitrogen use	Topsoil losses	Total nitrogen uses	Net revenue[b]
	thousand tonnes	*thousand kilograms*	*US$000*
No special controls	1.7	5.7	22.1
Special controls of			
Soil erosion[a]	0.4	6.7	20.2
Soil erosion[a] and nitrogen use	0.4	4.2	19.9

Source: Extracted from Nelson and Seitz (1979).
[a]Based on practices reducing losses to not more than the maximum Soil Conservation Service limits.
[b]A discount rate of 5 per cent was applied to future years' net revenues.

19.4.8   Policy and administration

Our brief survey shows that public financial policies should be based on wide understanding of socioeconomic systems, of the appropriate functions of government within them, and of the economics of farm firms. Social wants should be assessed and predicted by appropriate political systems. Technological and economic opportunities should be well recognized. Policies should then be formulated that are acceptable as reducing uncertainties and providing continuing opportunities to secure the desired production, savings, and investments, including the build-up of human skills and attitudes, for the future (Chapters 20 and 21).

Within countries, inter-regional differences should be well recognized, and social goals for the less favoured areas well chosen. A related noteworthy requirement of importance to agriculture is, however, that policy and administration should be based only on the general social interest. Partiality tends to result in misuse of resources, instabilities, and uncertainties.

19.5 STATISTICS

Many types of decision by governments and firms cannot be well made if reliable statistics are not promptly and widely available, in appropriate form. Thus modern agriculture depends heavily on statistics as part of the infrastructure for production, marketing, international trade, consumer relations, government policy formulation, and other matters. To secure statistics, government services are commonly essential, as well as co-operation by private firms and consumers. But many developing countries have great difficulties because of mixed cropping of small, ill-defined plots and fields, multitudes of small firms, lack of animal housing, much self-subsidence production, few 'bottle-necks' through which produce is marketed, and other obstacles. Even 'centrally planned' economies, such as China's, have severe difficulties in attempting to secure adequate statistics.

REFERENCES

Bastiat, F. (1850). *The Law* (translated from the French), Foundation for Economic Education, Irvington-on-Hudson, NY.

Clark, C., and Haswell, M. R. (1967). *The Economics of Subsistence Agriculture*, 3rd edn, Macmillan, London.

Nelson, M. C., and Seitz, W. D. (1979). An economic analysis of soil erosion control in a watershed representing Corn Belt conditions. *North Central Journal of Agricultural Economics*, **1** (2).

Scottish Office (1981). *Scottish Digest of Statistics, 1979–80*, HMSO, London.

FURTHER READING

Bock, C. A., and Myers, D. A. (1980). Taxation: federal, state and local systems. Report no. AE-4490, Department of Agricultural Economics, College of Agriculture, Urbana-Champaign, Ill.

Commonwealth Agricultural Bureaux (quarterly). *Rural Development Abstracts*, Commonwealth Agricultural Bureaux, Farnham Royal, Slough.

*Cairncross, S., Carruthers, I., Curtis, D., Feachem, R., Bradley, D., and Baldwin, G. (1980). *Evaluation for Village Water Supply Planning*, John Wiley, Chichester.

Clark, C., and Haswell, M. (1970). *The Economics of Subsistence Agriculture* 4th edn, Macmillan, London.

Coombs, P. M. (ed.) (1980). *Meeting the Basic Needs of the Rural Poor: The Integrated Community Based Approach*, Pergamon Press, New York and Oxford.

*Grant, W. W., and Daal, D. C. (1978). *Bibliography of Agricultural and Food Law 1960–78*, Minnesota Agricultural Experiment Station, Bulletin No. 523, Minnesota Agricultural Experiment Station, Minneapolis.

Gregory, M., and Parrish, M. (1980). *Essential Law for Land Owners and Farmers*, Granada Publishing, St Albans.

*Hanson, J. L. (1978). *Monetary Theory and Practice*, 6th edn, Macdonald and Evans, London.

*Hunt, K. E. (1969). *Agricultural Statistics for Developing Countries*, Agricultural Economics Institute, Oxford.

*National Farmers Union (1979). *Tax on the Farm: A Handbook for Owners and Tenants*, 2nd edn, National Farmers Union, London.

Padfield, C. F. (1979). *Law Made Simple*, 6th edn, W. H. Allen, London.

265

Prest, A. R. (1974). *Public Finance in Theory and Practice*, Weidenfeld and Nicolson, London.
*Tweeten, L., and Griffin, S. (1976). *General Inflation and the Farming Economy*, Oklahoma Agricultural Experiment Station Research Bulletin No. B-732, Oklahoma Agricultural Experiment Station, Stillwater, Okla.
US Department of Agriculture (periodical). Statistical publications, including the annual *Agricultural Statistics* and commodity and factor *Outlook* statements.

QUESTIONS AND EXERCISES

1. During the last 10 years, in the region you know best, what changes have been made in the infrastructure that have affected farm families? Explain briefly what the effects have been, and how they have resulted.
2. In the region you know best, what proposals are now discussed for changes in the infrastructure as affecting agricultural production and incomes? Explain the probable effects.
3. Find and tabulate annual index numbers of prices for the main farm products and factors during the last 15 years, in the nation or region you know best.
4. List the sources of current published statistics that would be useful in relation to farm production and farm product marketing in the area you know best.

Chapter 20

Infrastructure II: Education and research

20.1 INTRODUCTION

20.1.1 Position in the socioeconomic system

This is the last 'box' of Figure 13.1 to be considered in Part V. We have already noted briefly that it provides knowledge (information and skills), and so affects decisions in the management of farm and other firms, and in government. It makes contributions to the closely related 'box' of 'cultural values and knowledge', and this too affects these decisions (e.g. the values in Table 12.1). Satisfactions of a less durable nature are also derived from some educational activities, and by individual research workers and students from particular research activities.

20.1.2 Importance

Even in self-sufficient tribal cultures, much training is given in 'bringing-up' the young. In modern industrialized societies much more education is needed. Agriculture provides many examples of the importance of the build-up of information and skills, and their transfer. For example, the life cycle of an individual farmer allows only some 30–40 years of control of a farm firm, and up to about 12 years previous experience of some of the many biological, work, farm-economic, and socioeconomic variables in farm management. But during the whole period of possibly 52 years many changes will be experienced in these many variables. Reliance could be placed only on the individual farmer's own accumulation and use of experiences. But this would not be efficient. Even by the end of the 52 years, he would have much to learn that would have been useful during many previous years. Elsewhere, much knowledge and understanding relevant to his experiences and decisions exists, and new knowledge will be added during the 52 years. Therefore, much in agriculture depends on the assembly and storage of knowledge, its retrieval and analysis, its transfer to other users, and its re-arrangement for particular purposes. Much depends on the abilities of individuals and groups to play their own parts in these processes, and to use knowledge well in their own decision making.

Less widely recognized is that research and education are, and always have been, essential to those technical, economic, and social changes that will improve human welfare. For these changes depend on *new* decisions.

Within farm and other firms, governments, and family and other social groups, experiences provide 'feedback' knowledge and alter attitudes. So some knowledge is discovered in this way, and how to use it is learnt. But such education and research is best considered as within the education and research 'box' in Figure 13.1. Therefore, it is only from this 'box' that useful *change* can stem.

One important reason for considering all sources of education and knowledge in one 'box' is that most decisions are influenced by a variety of sources of education and knowledge, and the quality of the decisions depends on more than one type of source.

We have noted the results of wide variations in the quality of the decisions of farm firms (e.g. in Sections 11.10, 12.1, 16.2.3, and 16.3.3). It is thus evident that the opportunities to improve decisions are substantial where agriculture is already largely 'science-based'. Elsewhere, even greater opportunities may exist, and certainly seem to be demanded because of population and other social pressures.

We have also noted the nature of these qualities in management and labour that are needed for the future (Sections 16.2 and 16.3). These require, from both education and research activities, investments in what has been called 'human capital'. And further, we have noted the importance of education in (a) helping youth to choose wisely amongst different types of occupation and in (b) fostering attitudes, and knowledge for appropriate mobility between types of occupation, and between rural and urban areas (Sections 1.9 and 16.3.2).

Consideration in Chapters 14–19 of other parts of the socioeconomic system has also shown the importance of adequate knowledge and its rational use. Thus, for example, the distribution of rights in land, the methods of rewarding labour and management, the terms of loans, the provision of communication services – none of these can be satisfactory unless many matters are understood by sellers, buyers, lenders, borrowers, tax-payers, and governments and their agencies. The problems that arise from rapid increases in human populations, industrialization and urbanization, specialization, and inter-regional and international trade – all these also require increasing knowledge, and not only physical facilities for communication but also understanding, empathy, foresight, and rational conclusions. Likewise in many countries, attempts to secure conservation of natural resources, control of pollution, rural recreation facilities, and wise transfers of land to non-agricultural uses cannot for long be useful unless they are based on facts, understanding, and rational thinking. Wise judgements of these matters also require sufficient social cohesion, which depends on education.

20.1.3 Content

What priorities should be in education and research has long raised many questions. The answers given have depended on the patterns of values in particular cultures, included the values placed on change. Early in history, high values were placed on religion, military security, social cohesion, and law. The enquiries fostered were few and largely related to these values. Much knowledge was secured (Chapters 6 and 7), but only slowly and unevenly. Scientific methods, and years of formal schooling for all, began late in the human story, and even today are far from fully developed. The priorities therefore still differ widely in different countries and confusion and conflict arise between different social groups according to their valuations of, for example, (i) science and technology as against studies of human relationships and behaviour and the arts; (ii) mental as against physical health and abilities; (iii) existing as against changed governmental, bureaucratic, and professional structures; (iv) urban as against rural development; (v) efficiency as against equity; (vi) beauty; (vii) truth for use as against truth for its own sake.

For our purposes here, we can usefully consider education and research as contributing to *decisions* and to the carrying of decisions into effect. We are concerned therefore with knowledge (both information and skills). But we are concerned also with the determinants in young and old, of values and goals, and related attitudes and motivations. These affect all agriculture, because they affect (a) demands for outputs (e.g. through human population changes; ideas about what the consumption levels and patterns should be); and (b) all the inputs used, because human decisions sufficiently determine 'input mixes' (Sections 11.3–11.7 and 11.10). Even within the constraints of tropical and other climates on human physical work (Section 7.5), different individuals and groups can differ widely in the labour efforts they exert. What capital is used depends heavily on human decisions about saving and investment. How land and water are conserved and rights in them distributed are human decisions. Methods and levels of taxation are perhaps even more clearly human decisions. Therefore the 'qualities' of management have fundamental importance – in individual firms and social groups and governments, both local and central.

Such considerations emphasize further that education and research should foster the development and spread not only of knowledge but also of all those personal and social attributes that can ensure its wise use. These attributes include (i) those helping to secure *information* useful in improving the quality of decisions (e.g. prompt observations of biotic factors on particular sites; discerning observations in farm-economic planning); (ii) those helping to secure rational and timely *decision-making methods* in planning and the execution of plans; (iii) those concerned with '*foresight*', reducing risks and uncertainties, but also raising the willingness to bear appropriate weights of these; (iv) those securing within firms and the socioeconomic system as a whole those *economic and social relationships* that maximize human welfare (e.g. those concerned with land, labour, capital, markets, taxes, education itself).

Listed in another way, the attributes are many, including (i) the will and energy to be observant, and retentive, and to study variations and the causes of variations both natural and human; (ii) the will to forgo present benefits to secure future benefits or reduce future losses or costs; (iii) patience, perseverance, fortitude, but also foresight and flexibility; (iv) willingness and energy to think through problems independently but also to co-operate with other individuals and groups; (v) empathy with others and desire for their welfare and easy communication with them; (vi) respect for the morals, laws, and religions of particular societies but willingness to recognize and accept any changes that are inevitable or desirable; (vii) willingness to face unpopularity and resentments at times; (viii) willingness to be taxed so that infrastructures can be adequate and income distributions socially acceptable; (ix) honesty; (x) the stability and confidence that result from sufficient knowledge of history and past achievements.

So far, we have not considered research and education separately. When they are so considered, the purposes of research in improving the quality of education, and so the quality of economic decisions, may sometimes in practice become subsidiary purposes. Personal satisfaction from the pursuit of knowledge for its own sake and from social status as a discoverer, and realization of ideals about the freedom of the individual human spirit, may become the principal purposes. We cannot here consider the benefits and costs of the resulting different priorities in research, but must confine ourselves to research that probably will, within a reasonable period, improve education and decisions.

20.2 AGRICULTURAL EDUCATION, INCLUDING EXTENSION

20.2.1 Definitions

Many organizational and financial reasons result in 'agricultural education' being often considered separately from 'general education' as well as from 'agricultural research'. And within 'agricultural education', formal, and largely indoor work at few centres over many months and leading to examinations and qualifications is often distinguished from extension work – in shorter periods, dispersed, and often with older age groups and their local problems. But while such distinctions may be administratively convenient, fundamental purposes and the interconnections they require should be well borne in mind. Thus a youth who decides to enter agriculture should have acquired much from general education that will help him in his agricultural education, both central and 'extended'. And, in these and agricultural experiences, he should secure much that will add to his general education for family and other social decision making. And agricultural research should contribute to general education as well as agricultural education. The efficacy of extension work depends heavily on relevant research, the general education of farmers and farm households, and the general and agricultural education of the teachers or discussion leaders. The flows of information and skills, ideas and values should not be restricted by organizational barriers.

Nor should they be restricted as between special educational institutions and family or other sources. Table 20.1 lists many of the various sources of education

Table 20.1   Major sources of education.

Families and communities	Formal courses	Associations	Other
Families	Primary schools	Church	Apprenticeships (in-service training)
Relatives, other	Secondary schools	Agricultural improvement societies	
Song, Dance	Tertiary Colleges Universities		Observations, experiences
Drama, Art		Unions	
Sports	Further education	Co-operative Societies	TV, radio, other media
Follow-the-leader	Day release		
	Block release Occasional Correspondence, radio, TV	Smaller groups	Libraries
			Banks
			Buyers, sellers
			Consultants
			Health services
			Veterinarians

affecting agriculture. This shows that special institutions are essential to adequate results, but they provide only parts of the total education.

Extension education is often very closely interwoven with advisory or consultancy work. When farmers are advised, the advice may be educational in that it improves their store of information and skills, and even alters their goals. Or the advice may have only short term effects, simply as a type of management service input that will in future be repeated from outside the farm firm. Management services should, in some circumstances, be bought by farm firms, and even supplied free if they are necessary as introduction to certain types of education, or for certain types of research. But in general, the additional investments to ensure that consultancy work is educational seem economic, because they are comparatively small, and the future benefits are substantial. This is particularly important now in the poorer countries using consultants from abroad. But it remains important elsewhere (e.g. in farm planning; see Section 12.4).

20.2.2   Benefits, costs, and efficiency

The costs of education are so great that the relation of benefits to costs has been increasingly studied in recent decades.

Benefits   Inevitably benefits are difficult to measure because (i) they affect socioeconomic systems in many direct and indirect ways; (ii) educational processes and the flow of benefits are dynamic, so that investments now will have effects well into the future; (iii) some benefits should be assessed by social, aesthetic, and intrinsic values rather than by instrumental values which are more measurable (Sections 12.1 and 21.8); (iv) many of the confusions and conflicts over priorities in education as a whole relate also to agricultural education.

One type of assessment of the benefits of education and research can be based on the number of economically and socially successful innovations in agricultural production, and the rates at which they were adopted. Figure 20.1 gives some examples for the Aberdeen area, showing the S-shaped adoption curves. Farm firms can be classified according to how early or how late they adopted a new technique. The slopes of the curves indicate how able and willing the population of firms as a whole was to make changes. The number of innovations in a comprehensive review would indicate the relevance and success of research. Of course, other determinants should be borne in mind, including all those affecting management decisions, such as prices, subsidies, taxes, land tenure, net worths, loans (Chapters 12 and 14–19). We also should note that the slope of adoption curves can be determined partly by 'follow-the-leader' behaviour, and in some circumstances even by the commands of governments.

Costs The staffing and other costs of schools, colleges, extension services, and other bodies for education are accounted for, but the costs to families and individuals are difficult to measure. They include the personal efforts put into

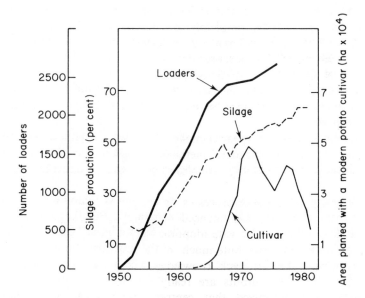

Figure 20.1 Contrasts in rates of adoption: hydraulic tractor-mounted loaders in Aberdeenshire; grass area cut for silage in Aberdeenshire area as a percentage of total grass area cut; area planted with a modern potato cultivar in Great Britain.

study, and the foregoing of opportunities in alternative activities. Thus in poor countries, the contributions of the labour of many farm children to subsistence are not forgone because they are regarded as more essential than schooling. And in almost all countries, many farmers' sons are not spared from farm work to attend college courses, running over two or three years. And everywhere personal valuations of efforts put into study vary widely.

Efficiency   How fully the objectives of agricultural education are achieved and with what efficiency depend on many variables. These include: (a) choice of the objectives themselves, their relevance to the economic and social circumstances, and their balance; (b) levels of innate intelligence; (c) existing social and personal values as affecting the desire for education; (d) organization and methods, and equipment, considering the whole range of sources (Table 20.1); (e) the detailed content as affected by assembly of relevant information and skills, including those from research; (f) securing and using feedback from studies of benefits as against costs.

20.3   AGRICULTURAL RESEARCH, INCLUDING 'DEVELOPMENT'

20.3.1   Definitions

Research and education are closely connected because both aim to secure knowledge, to use it, and to pass it on for use. In research, the emphasis is on the securing. And this includes completely new discoveries and assembling existing knowledge, relating it to different circumstances, and combining it into new concepts and systems. The purposes are essentially (i) understanding, and (ii) knowledge of quantitative relationships so that, for example, output–input relations in crop production on particular sites can be more reliably forecast, better machines can be designed, or supplies or prices better forecast and farm plans improved.

Such a wide definition of research implies that research methods include a wide range from rigorous experimental methods to case studies, and surveys, and historical studies covering many variables none of which is under the researcher's control.

'Development' of products, processes, or systems is commonly considered as different from 'research', and financed separately. But the stages from first concept to final suitability for wide adoption cannot be well classified in a simple way. Farm firms may contribute much to the first conception of hypotheses. They may provide early 'feedback' knowledge. They may test out research results years before the final outputs are ready for wide adoption. Increasingly researchers have recognized that central experimental work should be complemented by surveys, dispersed trials, and case studies. This is true even when the immediate purposes seem to be only within the biological or the work sub-systems of agriculture.

20.3.2 Benefits, costs, and efficiency

Benefits These depend on all the determinants of choice of priorities, as influenced by study of potential benefits, both long term and short term. These determinants include therefore knowledge of (a) the agriculture of the area considered, and the particular socioeconomic circumstances; (b) the natural and social sciences that could contribute. Benefits also depend on (c) education, both to contribute research skills and motivations, and to use and disseminate research results.

Many governments now recognize that the potential benefits are high although their full range cannot be well measured.

Costs To most governments, costs also appear high. Completely new discoveries are rare, and the staffing and other facilities that directly or indirectly help to secure them are costly. But even more important, the variations in climates, soils, and biotic factors, and in work, farm-economic research, and socioeconomic conditions inevitably require much research that cannot be centralized. And the 'subject matter' requires both specialists and general practitioners, and much equipment and travel. Even the assembly and digestion for a particular country of relevant research results from the world's output entails substantial costs. Moreover, where agriculture is already based largely on past research results, the costs of useful, new results may be high because research inevitably tends to have diminishing marginal productivity (Section 12.4.1). The productivity of particular types of research may of course remain high in sustaining existing levels of production efficiency on farms (e.g. by control of biotic factors). And there may be great research productivity from useful new discoveries. In many countries that have used research little in the past, high marginal products from research could still be achieved.

Efficiency As in education, the number of determinants of efficiency in research is great. Much depends on clear conceptions of purposes and probabilities, and on wise decision about priorities. The good timing of purposes, and the balancing of efforts for productive and richer areas as against those for remoter and poor areas are often especially important. And always imagination is required if probabilities and possible uses are to be well conceived.

Much depends also on methods, and the 'input mix' of staffing, equipment, travel, libraries, and communication facilities. Wise use and combination of methods requires some very scarce skills. Frequently, the personal and other costs of very 'difficult' work naturally seem especially high, but some of this yields more useful results than 'easier' work that is more repetitive. On the other hand, 'difficult' and costly research on ill-chosen subjects may be much less productive than research that wisely uses established principles and methods in biological, work, farm-economic, and socioeconomic systems and circumstances where they have not been well used before.

274

FURTHER READING

†Arndt, T. H., Dalrymple, D. G., and Ruttan, V. M. (Editors) (1977). *Resource Allocation and Productivity in National and International Agricultural Research*, University of Minnesota Press, Minneapolis.
*Binswanger, A. P., and Ruttan, V. W. (1978). *Induced Innovations Technology, Institutions and Development*, Johns Hopkins University Press, Baltimore.
*Brown, L. A. (1981). *Innovation Diffusion: A New Perspective*, Methuen, London.
*Crough, B. R., and Chamala, S. (1981). *Extension Education and Rural Development*, Vol. 1, John Wiley, New York.
Dillon, J. I., and Hardaker, J. B. (1980). *Farm management research for small farmer development*, FAO Services Bulletin X 1041, Food and Agriculture Organization, Rome.
†Evenson, R., and Kislev, Y. (1975). *Agricultural Research and Productivity*, Yale University Press, New Haven, Conn.
Food and Agriculture Organization (1977). *Books and Periodicals for Agricultural Education and Training in Africa*, 2nd edn, FAO, Rome.
Food and Agriculture Organization (1980). *Agricultural Training: Report of an Evaluation Study*, FAO, Rome.
Hirst, P. H., and Peters, R. S. (1975). *The Logic of Education*, Routledge and Kegan Paul, London.
†McAnany, E. G. (ed.) (1980). *Communications in the Third World: The Role of Information in Development*, Praegar, New York.
McClay, D. R. (1978). *Identifying and Validating Essential Competences Needed for Entry and Advancement in Major Agricultural and Agribusiness Occupations*, US Government Printing Office, Washington, DC.
*Mosher, A. T. (1978). *An Introduction to Agricultural Extension*, Agricultural Development Council, New York.
†Peterson, W., and Hayami, Y. (1977). 'Technical change in agriculture'. In *Survey of Agricultural Economics Literature*, Vol. 1 (L. R. Martin, ed.), North Central Publishing (for the American Agricultural Economics Association), St Paul, Minn.

QUESTIONS AND EXERCISES

1. List and briefly explain what you consider should be the main purposes in education of rural children up to about 21 years old.
2. Plot statistics showing the adoption of one innovation by farmers.
3. For the rural area you know best, which of the sources of education listed in Table 20.1 do you believe are not important. Which should be used more and developed further? What other important sources if any, should be added to the list?
4. Give three examples of agricultural problems that you judge should have higher priority in research programmes. Explain your reasons.
5. For farmers in the area you know best, list the ways by which (a) they can influence research priorities; (b) they can secure research results and have explained the relevances to their production and marketing.

Part VI

Introduction to policies

Chapter 21

Concepts of development

21.1 INTRODUCTION

The understandings that we have secured in Parts I–V can now be used to obtain a basic understanding of the development of agriculture and the objectives of agricultural policies. This will help when we consider 'Organization' in Chapter 22 and 'Some successes and failures' in Chapter 23.

21.2 A DEFINITION

'Development' is now a common word, but it has various meanings. In plant and animal development, the processes are biochemical. Socioeconomic processes are essentially different: they depend on human will and decisions in the face of both natural and human conditions.

In relation to agriculture and socioeconomic systems, we can best understand 'development' as one among a range of words including 'change', 'growth', 'progress', 'advance', 'modernization', and 'reconstruction'. Perhaps the commonest meaning of '*development*' is 'those socioeconomic changes that result in a sustained increase in the annual output of goods and services per head of population, which increase is called "growth" '. But many of those concerned would insist that 'development' should not only raise average incomes but also 'improve' income distributions within and between the nations. And those with yet other goals in mind often speak of 'development' when they mean 'progress' or 'advance' towards *these* goals, including perhaps 'modernization' by the copying of production and marketing methods in a richer country. Thus judgements are made of what changes would be favourable for human welfare, and 'development' implies that these changes are achieved in sufficient measure.

In real life, however, changes can occur that *reduce* the total goods and services available for consumption in particular groups in a population, or per head of the population as a whole (e.g. rapid population increases; soil depletion, and erosion; increases in weeds, pests, and diseases; debilities in the farm labour force; wars; failure of resettlement schemes; unfavourable terms of trade for exports as against imports). In individual farm firms, changes may be unfavourable rather than favourable (Table 18.1), yet the 'process of unfolding or unrolling' is there just as much as in all agriculture or in national economies.

Moreover, if this unfolding process is called 'development' only when it has particular results that in themselves seem favourable, then the costs or burdens of accompanying changes may too often be neglected.

For our purposes therefore, the most useful definition of *'development of agriculture'* is 'changes in the biological, work, and farm-economic sub-systems of agriculture, and in their relations with the socioeconomic system, that are of more than short duration and have more than short-term consequences; and the processes that affect these changes or determine their consequences'. Thus some of the changes may prove favourable and some unfavourable. Some now may determine others years later. Some may be fostered or imposed from outside agriculture. Some may be natural, rather than the direct results of human decisions. Particular social groups in or outside agriculture may play particular parts in bringing about change. Different groups may be affected differently. The changes may be accelerated or slowed up, or halted. The definition can be used in all countries, and does not depend on any ideas of 'under-development' or poverty. The definition requires logically that those who approve some particular development should state the criteria by which they judge.

21.3 DEVELOPMENT IN THE BIOLOGICAL SUB-SYSTEM

In Chapter 1 and Parts II and III, many indications were given of important developments. Some started with man's observations within his old territories – e.g. domestication of animals, cultivation possibilities of wheat, selection of other crop plants, irrigation possibilities, better timing of sowings, how to control weeds, how to improve storage of grain. Other changes depended more on observations as man pioneered in new territories. Climatic changes and wars caused unfavourable developments in many periods, but in general *Homo sapiens* had biological success as measured by population increases. This success itself, however, caused Malthus and others to fear yet more unfavourable developments.

In the era of 'science with practice' (Section 6.4) fears were dispelled. Science explained many biological opportunities, and measured them in terms of output–input relationships at particular sites, or in particular animal populations and environments. But even such discoveries as the importance of nitrogen or phosphate to crop plant growth are not yet fully used even in western Europe or North America. And indeed, when we look closely at any one area, we can see too that even the skills and information of the types that 'practice' has afforded are not fully used by many farms (e.g. in seedbed preparation, timeliness of sowing, weed control, cow management). Moreover, the newer contributions of science are many and the development resulting from them is far from complete (e.g. chemical and biological controls of biotic factors; greater use of nitrogen fertilizers in Illinois and Saskatchewan (Figure 11.6); the plant breeding, irrigation, and fertilizer studies in the 'green revolution' in Asia). And 'practice' too is still yielding knowledge which is very seldom widely used immediately (e.g. about grass silage handling; cow management; minimum cultivations). Thus although biological developments have been great everywhere, and specially great where 'science' and 'practice' have been well married, we have seen that they are still uneven as between farms and as between areas.

These basic facts suggest that because everywhere farmers differ in the decisions they take, there are everywhere opportunities to foster those types of development that are judged to be the most desirable. Even in the agriculture of poor countries with few or no *new* biological inputs (such as hybrid seeds, inorganic fertilizers, or other agrochemicals) total useful production could in theory be raised in many areas by some 50 per cent, *if* all farmers and their labour forces copied *intimately* the input mixes and timings of the farmers with highest production (Haswell, 1953).

The within- and between-area variations also point to the importance of our understanding of 'accelerators', 'brakes', or 'obstacles'. Why was development in some areas and countries so much more rapid than elsewhere? Why was the process of raising outputs per head of population not sustained in many countries with much poverty (Section 1.5)?

Such questions lead us back to decision making on individual farm firms. But before re-considering this, we should recognize again the importance of developments in the other sub-systems and the socioeconomic system.

21.4 DEVELOPMENT IN THE WORK SUB-SYSTEM

Changes here can sustain and accelerate biological changes. For example, early observation of the benefits of irrigation and careful grain storage led to the *sharuf* and underground stores (Section 7.3). The possibilities of the use of draught animals for tillage and the use of metal tools for tree and bush felling led to major land-use changes. Inventions of reaper blades, binders, and combine harvesters sustained the pace of settlement of the North American prairies for grain, and developments in livestock production in western Europe. Tractors have played a major part in securing timeliness, and therefore higher crop yields (Section 11.3.3).

Some inventions of equipment can be judged to have induced biological changes. (For example, the harnessing of wind for pumping water and ideas for all the surveying and layouts required for the control of water in the Netherlands seem to have been at the very conception of biological developments there. The designs of harvesting machines have led plant breeders to alter their definitions of the 'ideal' cultivars of grains, cotton, and other crops (Section 4.3).)

Again, we can recognize that 'science and practice' made stimulating contributions of new information and skills. These have increased rapidly since about 1890 (Section 7.3). But again we can see that the resulting developments on farms have not been uniform, so that there are substantial variations in the productivity of labour and machine use within areas and between areas (Tables 9.5 and 11.1). So again we are led back to decision making in farm firms, and to consider what induces changes, what determines their pace, and what stops them.

21.5 DEVELOPMENT IN THE FARM-ECONOMIC SUB-SYSTEM

Where agricultural production is largely for consumption by farm households themselves, changes in population can easily be recognized as the inducers of other changes. Increased numbers *demand* more food, fuel, and shelter, and *supply* more labour. So more land can be tilled, fallows and uncultivated areas reduced, and crop yields increased, or reduced by over-cropping which leads to shifting cultivation or migration. But even such conditions and their challenges result in significant differences between households in their decisions, as reflected, for example, in their production per head. In close studies of particular areas, differences of 80 per cent or more have been found between the least production per head and the most (Haswell, 1953; Barlow *et al.*, 1983).

The next obvious inducers of change are opportunities to sell products and buy goods for consumption – non-agricultural items, or perhaps especially attractive items from farms in different climatic and soil conditions. Such inducers existed in prehistoric times for there was trade in salt, metals, precious stones, cloth, kola nuts, and many other items. Governments required 'revenues' in the form of produce or labour so that taxation also induced changes. Wars resulted in plunder and more taxation, but often also led to trade. In modern times, the possibilities of trading have been greatly increased and many new factors have become available. We noted that the farm production–market flows in our diagram (Figure 13.1) can usefully be regarded as the *axis* of the socioeconomic system. All the many determinants of supplies, demands, and prices in markets for products, and of supplies and prices in markets for land, labour, machinery, loans, and other factors, can induce changes in farm-economic systems.

Moreover, governments can cause changes over time in the farm economies of particular areas as well as wide differences, between areas, because they can supply services through infrastructures and education and research, tax and subsidise, and intervene in product and factor markets (Figure 13.1 and Part V).

Thus there are now, in virtually all countries, many social and economic influences that can induce and affect changes in the farm-economic sub-system. But again, the final results depend on the decisions of farm firms, and the variations in these within areas as well as between areas are highly significant. Even as between collective farms in the Ukraine in the USSR, there are substantial differences in productivity and net returns per worker (Mikitenko, 1978).

21.6 DEVELOPMENT IN THE SOCIOECONOMIC SYSTEM AS AFFECTING AGRICULTURE

Increases in population, the opening up of product and factor markets, and the government activities just mentioned are not the only socioeconomic changes that can influence development on farms and elsewhere. Changes in ideas, knowledge, and values in the 'cultural values and knowledge' box of Figure 13.1 result from the culture contacts that trade and communications foster. Many other types of education lead to readjustments of ideas and values. Urbanization and industrialization lead to reassessments by individual families, as well as by governments, so that decisions about consumption and about production, conservation, and recreation are altered. Changes in consumers' incomes, and their distribution, lead to changes in consumers' decisions and so affect much in agriculture and the socioeconomic system. Governments do not fully control such changes,, nor yet the instability of general price levels (Section 19.4.4). Experiences of old land tenure and loan arrangements can lead to changes in farm decisions and in factor markets and the infrastructure, and some of these changes may not be controlled by governments.

Again the opportunities for change and the variety of influences have resulted not in uniformity but in significant variety. Even within western Europe many socioeconomic differences have resulted because decisions have differed. And the same is true within India, and even within the USSR and China.

21.7   AN OVERALL VIEW

This brief review of development shows again how closely interrelated are the biological, work, and farm-economic sub-systems to one another and to the socioeconomic system. Changes in any one soon have consequences in the others (see also Section 2.3). So the *starters* of developments that are judged to be worthwhile can be in various parts of the whole system of agriculture. And likewise, the *accelerators* can be in any part, or in more than one part. The connections can have advantages because they make it possible to speed worthwhile developments in various ways. And these can be co-operant so that the results are more favourable. But both Nature and Man have various *brakes* in the system and *obstacles* that are difficult to avoid.

Our brief review has also indicated the basic importance of decision making by farm firms, by others in markets, and by governments. All this decision making can be considered as determined by (a) values and goals; (b) challenges of problems; (c) knowledge (both information and skill); (d) reasoning; (e) other resources; (f) the obstacles and brakes just mentioned. Because all of these can differ within cultures, between social groups, and between individuals, decisions differ and, as we have noted, results are not uniform even within small areas.

But development is change and the causes of change. So the *changes* in (a)–(f) may be more important than their states at any one point of time. Whether the obstacles and brakes of (f) change may of course be especially important, and the natural endowments in (e) are fixed.

We should also understand that (a)–(f) are all interrelated. Thus values may be such that problems are recognized earlier, new knowledge is highly valued, strict objectivity and logic in reasoning out policies and in management are respected, useful new chemicals and machines are quickly adopted, obstacles to increased production, marketing, and consumption are removed. And so on. Or values and education may be such as have the opposite effects. Figure 21.1 summarizes briefly the interrelationships in a diagram. In real life, each of the boxes is of

282

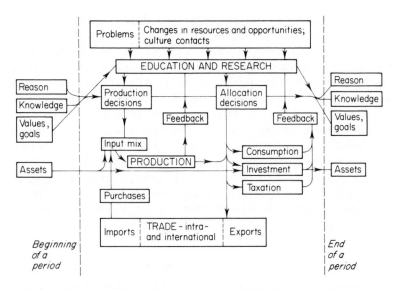

Figure 21.1 Simple diagram of some important relationships in development.

course much more complex than can be indicated; and the processes are continuous. Obstacles and brakes of many kinds affect the flows and boxes but are not shown. Even so, such a simple diagram can be a basic help when we think about development.

The definitions of knowledge, education, and research for Figure 21.1 are the wide definitions that we have already used. We need here only to note again (a) the fundamental contributions that the whole development system can obtain from education and research; (b) the various contributions *to* them from (i) problems; (ii) culture contacts, trade, new resources and opportunities; (iii) the results of decisions (experiences in management and labour); (iv) consumption experiences; (v) investment.

21.8 VALUES AND GOALS

21.8.1 Differences and conflicts of interest

The many goals affecting the decisions of farm firms are related to the values of individual firms (Table 12.1). These may be regarded as derived from education and research within the cultures and agricultural and socioeconomic conditions of particular areas and countries (Figure 21.1). Value patterns therefore differ from country to country and in lesser degree from area to area. But even within areas, between farm firms, the differences can be substantial, as we noted. Moreover, the value patterns of different groups in particular societies may differ so that their 'interests' seem to conflict. Thus, for example, many rich landowners in the UK place high values on high scales of living for their families; high levels of conservation and rural beauty; hunting, shooting, and fishing; some traditional farm production practices; and farm animals of especially attractive conformation. There are some conflicts within this value pattern itself but even more with the value patterns of some hired farm workers. They put high instrumental values on higher wages, and, so that these can be paid, on more intensive farming to increase the total net value added, but with the shares going as rents and returns to management reduced. Most urban consumers value highly adequate and secure food supplies of high quality and with ample scope for choice. These also require productive farming. But urban consumers place high values too on many other goods and services and on job opportunities that depend on exports of manufactured goods and services to countries from which foodstuffs could be imported. They therefore want cheaper products from UK farms and more food imports. Moreover, they value highly rights of access to rural areas that can conflict with intensive farming and the landowners' interests in conversation, beauty, hunting, shooting, and fishing.

Conflicts of interest can of course be quite direct and very deep because of instrumental values and related goals such as higher and more secure incomes. They are greatest where average incomes are low, the ownership of assets is very unequally distributed, and where there is 'imperfect competition' (Chapters 14, 15, 17, and 18). Thus, for example, conflicts over land-use rights and rents can be deep between landowners, tenants, and the landless in densely populated poor countries such as the Philippines, India, or Indonesia. The conflicts of interest between the unemployed and landless youth and the employed rich with land-use rights have increased greatly in western Europe and North America, although limited by social security benefits from governments.

21.8.2 Classification

Philosophers have made many attempts to classify values. An ancient classification which is still useful is into those concerned with (i) truth, (ii) beauty, and (iii) goodness. The human judgements of the values in all three groups are seen to be related to the enhancement of human life. Values are highest when the greatest contribution is made to the coherent functioning and organization of human experience as a whole. The groups, sometimes labelled 'logical', 'aesthetic', and 'ethical', are therefore intimately related, but they cannot well be reduced to one or two groups. The 'instrumental', 'expressive', and 'intrinsic' values that we considered in relation to Table 12.1 can be regarded as derived usually from more than one of the three basic groups. (For example, the 'intrinsic' values of working hard with traditional farming methods derive from judgements that are of some 'ethical' and 'aesthetic' values.)

21.8.3 Truth

One of Confucius' famous conclusions was that 'Truth does not depart from human nature'. He stressed the importance of human relations. His concern was with ethical values. Facts about climates, soils, plants, and animals were much less important. Statistics of prices, supplies, demands, wages, migrations, and interest rates would in Confucius' value pattern have been considered much more as poor reflections of human relationships and behaviour than as facts for economic analysis. But it is convenient to consider truth to be logic, and objective facts about Nature and human behaviour. We then need not be confused by ethical (goodness) values, although we should undoubtedly have these in mind when using truth.

We can see clearly that societies that value truth highly can develop differently and quicker than societies that do not. The observation, discovery, acceptance, and logical use of facts resulting from all the flows into and within 'education and research' can start, accelerate, and sustain many developments in all the sub-systems of agriculture and in the socioeconomic system as affecting agriculture. This will be clear if we consider Parts II–V and Chapter 20 from this standpoint. Formal education and institutionalized research are highly important, but so too are the valuations of truth as affecting all the activities of firms, families, groups, and governments.

The making of laws and their results are obvious examples, even though ethical and aesthetic values also contribute. The results arise of course from decisions about policies and organizations and plans, but also from the innumerable day to day decisions in management and labour. Thus how highly truth is valued can explain many of the differences we have seen between farm firms within areas, between areas, between countries, and over time.

Valuations of truth can be crucial when societies face the challenges of new ideas, organizations and methods from other cultures, as for example the Inca did in the fifteenth and sixteenth centuries, and Japan and China in the nineteenth.

Usually less dramatic but still very important is whether truth is valued highly when the conflicts of interest between social groups within a country are considered. Uncertainties and confusions are all the greater, and the dangers of conflicts and illogical decisions all the more serious, the less the facts are known and the less logical the reasoning. This is true even when some of the desirable facts cannot be obtained, and so *political* judgements must be made. Conflicts between groups in western Europe and North America are now substantial over ethical and aesthetic values (e.g. of conservation of wild life and landscape). Even such conflicts can, however, be reduced when truth is valued highly (e.g. in clear statements of the *probable* outcomes of alternative decisions).

Part of truth is about uncertainty itself and how to be logical when uncertainties are present (e.g. in farm planning; Section 12.4.5). Societies develop more satisfactorily if they value both (a) the narrowing of uncertainties by facts and logic and (b) logical responses to unforeseen changes (e.g. through (i) flexible decision making in markets for grain so that changes in storage, trade, and utilization reduce price fluctuations; (ii) revised birth control policies when medicine has reduced death rates).

Many conflicts in societies are in essence not about truth, beauty, or ethics, but about organization and the power to put political decisions into effect. Such conflicts are often symptoms that truth is not highly valued.

21.8.4 Beauty

Visual and other perceptions of the beauty of home territories probably have considerable effects on decisions such as whether to transfer from agriculture to urban employment, or migrate to other countries. The aesthetic values of known landscapes and wildlife also affect the decisions of farm firms (Section 12.1.1) and local and central government decisions about conservation (Sections 1.9, 19.5, and 19.6). Such decisions can reduce outputs of foodstuffs below what they otherwise would be, at least in the short run. They can include decisions about sizes of fields and farm layouts and buildings, and even farm sizes (Section 8.5). They can therefore affect labour efficiency and income distributions.

Other aesthetic values relate to the beauty of craftmanship and design, and of animals that have been well bred and reared. Where such values are very obvious in value patterns, they affect the development of agriculture (e.g. through retention of traditional buildings in the Federal Republic of Germany). Efficiency in securing output may of course seem 'beautiful', but essentially efficiency itself has ethical (goodness) values, and there are other values in design and craftmanship that are distinguishable from these.

Perceptions of the tastes and odours of foods and beverages also have values in the aesthetic class. Perceptions of succulence, tenderness, and other physical attributes must also be included. So these aesthetic values affect demands for foodstuffs and not least demands for quality in particular kinds. Thus in poorer, rural households heavily dependent on high proportions of cereals or starchy roots in their diets, choices of cultivar are adjusted to their flavours, final 'mealiness', and other attributes as experienced by the eaters. Hence the care now being taken to secure 'eating qualities' in high yielding rice varieties, potatoes, and yams. Many countries have long traditions of valuing eating qualities highly, because aesthetic values were expressed in farmers' and landowners' households able to arrange carefully for food supplies to accord with them. Hence the culinary and agricultural traditions of, for example, France and China. Where market pricing is largely used to register quality preferences (Section 13.3.1), price premia and discounts serve to indicate to producers and traders these aesthetic values to consumers, within the limits of their purchasing powers. But marketing arrangements for the assessments of qualities and price formulation may not give wholly reliable indications. Consumers may not have the range of possible qualities presented to them. They may be confused by packaging and labelling. The necessary supplies of the particular quality preferred may not be 'repeatable' enough to establish preferences. Therefore much development in production and marketing (and processing and cooking) depends on how far discrimination over eating qualities is sustained through education in families and elsewhere, and how far it is expressed through prices, but also in other ways.

Qualities in textile fibres, leather, and other non-food products also have aesthetic aspects.

Aesthetic valuations obviously differ between individuals, families, other social groupings, and nations. The differences help to explain differences in development and so add to the explanations contributed by differences in the valuations of truth.

21.8.5 Goodness

The many values in this third class are all those that are not put into the truth or beauty classes. When phrases such as 'the greatest good for the greatest number' are used it is often goodness values that are felt to contribute most to 'good'. From a biological standpoint, many are related to survival, reproduction, health, comfort, the upbringing of children, and the whole natural cycle of birth, life, and death. From a social standpoint, these matters are basic, but some goodness values relate to social behaviours and structural details that are not so directly concerned with biological matters (e.g. the status of bureaucrats, or landowners). Goodness values are closely interwoven with truth and beauty values and often co-operant with them in affecting decisions. But they should be distinguished from them.

Within the 'goodness' class, many values are 'social' in that they arise from relationships between individuals, between individuals and social groups, and between groups. The instrumental values placed on higher and more secure incomes are largely derived from social values of high and secure levels of consumption for families.

Social values have been derived – via education and research, as we defined them – from experiences during long prehistoric times and throughout history. Behaviours that these experiences seemed to make most satisfactory were adopted as 'folk ways' and in time as 'morals', and, later to substantial extents systematized (codified) into 'laws' (Section 19.2.1). Thus many of the values we noted as 'social' in Table 12.1 for individual farm firms are 'folk way' or 'moral' values. Many decisions in buying and selling, land tenure, labour relations, lending and paying back, taxation and the paying of taxes – indeed in the whole of the socioeconomic system (including government policy making and bureaucratic operations), depend fundamentally on 'morals' (e.g. when contracts are made; when laws are interpreted). Ethics is the branch of philosophy dealing with morals. Anthropology and sociology study morals in other ways.

As affecting development we can usefully consider another type of value within the goodness class. This type derives from judgements of relations of individual humans to Nature, and to any super-natural power with 'absolute' values to which all other values may be subordinate. Many of the struggles with Nature of pioneers, mountain sheep farmers, indeed of many other farmers too, can be recognized as providing values of this type. Many 'intrinsic' and 'expressive' values can be included (Table 12.1). And in the endeavours of *individuals* in research and education, in markets, backrooms, committee rooms, and elsewhere in socioeconomic systems, there often also seem to be values derived from 'dedications'. These are much conditioned by social experiences and education, but their sources seem to be deeper – from Nature and perhaps God as a super-natural power. Societies have long recognized these relationships of individuals, and formed morals about them. Religions are concerned with them as with other morals.

Differences in the goodness value patterns of different countries have determined, along with their other values, big differences in socioeconomic and other agricultural developments. Examples can be seen in the wide contrasts between Japan and China, New Zealand and Patagonia, and in more detail in our Tables 3.4, 3.5, 9.5, 10.1, 10.2, and 12.2. Current examples of the importance of differences in goodness values can be seen in many debates over world food supply problems, the 'population explosion', pollution, conservation, unemployment and the landless, and international trade.

Because the morals of a society are inevitably so important in its development, it is especially significant that morals were derived often painfully over long periods of the past and have determined religions, social structures, and laws. When the foundations of these were laid and during the periods when they were largely built on, societies' experiences were different from those they now face. But religions, social structures, and laws are not changed readily. Rather than evolve, they are liable to break down with serious consequences. The changes in death rates, population pressures, techniques of birth control, communications and culture contacts, wars, education and research, technological opportunities, urbanization, and international organization are all examples of how fast and how much the environments of societies have altered this century. They are therefore also examples of why moral values and the institutions built on them should be reconsidered and evolved. Without doubt, many very old morals and their reflections in social structures and laws are still valid indicators of the behaviours that will prove most satisfactory. But in many societies, new and supplementary morals are needed in the new circumstances. For example, where landless, unemployed populations in urban slums are increasing rapidly and many farms are already very small, the morals guiding decisions about divisions of labour, patronage, family responsibilities, and local charity need evolution into morals that can be effective about educational opportunities, birth control, enterprise and the provision of employment, taxation, land reforms, marketing arrangements, loans, meeting basic nutritional needs, extension education, etc. And the debates about agricultural prices and international trade and unemployment and security issues are all examples in many countries of the need to think through to satisfactory foundations for decisions and therefore to adequate morals.

Revisions of valuations of truth and beauty may also be needed to help in the face of many new challenges including, for example, needs for modern taxation and conservation.

21.8.6 Individuals, groups, and leaders

Although development is commonly studied using only a few national *aggregates* of economic data, and thinking of the socioeconomic systems of *nations*, the 'grass roots' of development are in individual firms and other operational units. The socioeconomic environments of these may be altered somewhat by government decisions (Chapters 14–19) but the problems and conflicts of interest of different social and economic groups, some of them quite small, are usually first considered by governments. So development depends not only on the values and goals of governments but also on those of firms and operating units and of social and economic groups. What all of these have of knowledge, reasoning, values, and education (Figure 21.1) determines which goals have priority. But because problems are seldom easy, the emergence of effective leaders is essential to wise decisions under all forms of government.

21.9 SOME IMPORTANT ACCELERATORS, BRAKES, AND OBSTACLES

21.9.1 Demands for incomes, and supplies of factors

The idea that 'each new baby brings its own rice' has some validity amongst families producing for their own consumption. Family labour inputs can later be higher and demand for the produce is assured. Likewise, in commercial agriculture changes in demands for income and willingness to supply the efforts to secure it are fundamental to development. Many biological, work, farm-economic, and socioeconomic conditions determine the relationships of incomes to efforts of labour and management and how variable such relationships are from year to year. But personal and farm-economic decisions have to be made about how much effort to supply. Such decisions are altered by changes in (i) the demand for incomes; (ii) the income–effort relationship; (iii) the reliability of results; (iv) the conditions for labour and the disutilities felt in efforts (e.g. in the Tropics; see Section 7.5); (v) the conditions for farm management (Section 12.6). Such changes in turn depend on all the flows into 'production decisions' in Figure 21.1. In all this, there can be 'accelerators', even 'starters' (e.g. ideas about the instrumental value of additional educational opportunities; research and education to improve crop yields; reduction of price fluctuations; new machines to ease labour; rational farm planning).

Contrariwise, there can also be various 'brakes'. Educational opportunities may seem to be inadequate and unlikely to be rewarded enough in future job opportunities. What extension education can offer as improved farming techniques may not be effective enough or feasible in farm economics. The natural endowments available may be inadequate. Governments may not be succeeding in reducing price fluctuations. No new machines may be suited to the particular work and farm-economic conditions. No help may be available in farm planning.

21.9.2 Wealth and income distributions

The values to some poor social groups of more equitable distributions of wealth (net worths) and incomes may rise fast, or be more effectively expressed. Governments and their bureaucracies are then led to formulate policies on reforms in land tenure and related provision of loans, extension education, and marketing facilities. Or such policies may result from the realization by 'government circles' in poor countries that rapid rural to urban migrations are endangering social structures and should be slowed. But the brakes on effective reforms may be lack of sufficient detailed knowledge and competent staff; social structures that give much power to landowners and moneylenders; goodness and truth values such that enterprising firms with high demands for incomes and new techniques can hinder or defeat the main aims of the reforms by fictious transfers of ownership, or by renting land.

21.9.3 Infrastructure and other socioeconomic services

When surveying markets for products and factors and the infrastructure (Chapters 13–20), we noted many changes that can affect the socioeconomic environment in which agricultural production and marketing decisions are taken. We can therefore understand that many types of decisions can accelerate, slow, or obstruct development through affecting the socioeconomic environment. Governments are seen to have great responsibilities over education and research, communications, land tenure, loans, price levels and interest rates, trade, security, taxation, subsidies, human migrations, conservation, and other matters. But their powers depend on the values and goals of social groups, firms, and individuals and their efforts day by day to give them effect. Therefore development can be accelerated by many types of decision if the values and goals of people and governments (including their bureaucracies) are in accord. But developments are slowed or stopped by lack of sufficient agreement.

21.9.4 Top national priorities, and political power seeking

Governments and peoples can of course be led to value highly national prestige and international status, or to give very high priority to the military defence of national security. And political groups, valuing their own power, may promote political ideologies more than rational solutions of current problems. So the changes in development can be determined by such values and the national and international conditions that can sustain them.

21.9.5 Attitudes to culture contacts, trading, and international finance

Another set of values that obviously determines development is that affecting decisions about culture contacts and new technologies. Those affecting decisions about foreign trade in goods and services and imports or exports of capital are closely related and also very important as affecting value patterns and economic assets, inputs, outputs, and allocation of outputs (Figure 21.1). Inevitably, development is most rapid where international contacts, trading, and finance are valued highly and not restricted. Hence, in large measure, the rapid developments of Japan, Taiwan, Hong Kong, and Singapore, as compared to those in much of mainland China and India. This is not to say that there is accord in people and governments about all the changes that resulted with either rapid or slow development.

21.10 PRINCIPAL CONCERNS OF ECONOMISTS

Most economists have recognized that no one basic principle can explain how development starts and how it may be sustained, accelerated, slowed down, or stopped. There is no accepted single theory of development. Nor have biologists, psychologists, philosophers, theologians, anthropologists, sociologists, or political scientists agreed such a theory. But we should note that the crucial importance now of development for human welfare has led to innumerable studies and debates and, at particular times and places, to concentrations on only one or two aspects.

After World War II, for many of the poorer countries, the priorities were (i) the ending of colonial governments and (ii) industrialization to provide urban employment and manufactured goods to substitute for imports. These priorities led to concern for the necessary investment and flows of capital and technology, and therefore international aid. Labour supplies and training were not emphasized, but 'take-off points' were envisaged where all was contrived to secure self-sustaining, future economic growth. Later there was recognition that goals for agriculture should not be neglected: investments and reforms to secure greater output from agriculture were seen to be essential. Population increases aggravated fears of food shortages and food aid from overseas was required. Emphasis was put on research and education, and the 'green revolution', and some on land reform. How far 'surpluses of labour' should be drawn from agriculture and rural areas and how development should be planned where big 'labour surpluses' were thought to exist was debated. Emphasis was placed by some studies on how past developments led to 'dualism' in that important sections of poor economies were based largely on imported technology, assets, and management. Much of the net value added by these sections was exported because the wages paid for indigenous labour were related to the poor alternative opportunities of this labour. Low wages and high social barriers were seen to reduce the flows of money, ideas, skills, information, goals, and values from these sections into the poor countries' basic economies. The 'trickle down' was small. Interests grew in 'balanced development' as well as in 'induced development'. In recent years, the World Bank and others have emphasized the importance of plans to reduce income disparities so that the poorest groups, especially those in the poorest countries, can at least have their 'basic needs'.

In richer countries, the concerns of economists have been much more with the acceptance of technological (biological and equipment) changes and their effects on agricultural productivity and marketing, including processing, storage, and international trade and aid. Many studies have been made of the determinants of changes in prices and incomes and of how to reduce year to year uncertainties about these as well as about biological yields. Grain and other surpluses and changing import deficits have led to many commodity studies and studies of international trading controls. In recent years, many studies have been related to the values placed by social groups and by governments on family-sized farms, conservation and pollution control, land-use planning, conditions of employment of hired workers, and aid to the poorer social groups and the less favoured areas.

A conclusion from the history of priorities in economic studies and 'government circles' is again that development cannot be fast, and its particular goals judged later as appropriate, unless the many components of the sub-systems of agriculture and of socioeconomic systems as affecting agriculture are *all* borne sufficiently in mind when studies are chosen and pursued, and when policies are formulated. Policies and the organization to give them effect must be well contrived to suit particular socioeconomic systems that have inevitably their own unique features.

REFERENCES

Barlow, C., Jayasuriya, S., and Price, E. C. (1983). *Evaluating Technology for New Farming Systems: Case Studies from Philippine Rice Farming*, International Rice Research Institute, Los Banos, Philippines.

Haswell, M. R. (1953). *Economics of Agriculture in a Savannah Village*, Colonial Research Studies No. 8, HMSO, London.

Mikitenko, I. A. (1978). Interrelations and development of labour productivity and remuneration [In Russian]. *Visnik Sil 's'kohgospodaurskoyi Nauky*, **5**, 102–106, Ukraine Research Institute of Agricultural Economics, Kiev.

FURTHER READING

*Anthony, K. R. M., Johnston, B. F., Jones, W. O., Uchendu, V. C. (1979). *Agricultural Change in Tropical Africa*, Cornell University Press, Ithaca, NY, and London.

Braider, J. (1980). *Prairie Farm Policy Guide, 1980–1*, The Western Producer, Saskatoon.

Commonwealth Agricultural Bureaux (monthly). *World Agricultural Economics and Rural Sociology Abstracts*, Commonwealth Agricultural Bureaux, Farnham Royal, Slough.

*Chambers, R. (1980). *Rural Poverty Unperceived: Problems and Remedies*, World Bank, Washington, DC.

Chou, M., and Harmon, D. P. (eds) (1978). *Critical Food Issues of the 1980s*, Pergamon Press, New York.

Cochrane, W. W. (1979). *The Development of American Agriculture: A Historical Analysis*, University of Minnesota Press, Minneapolis.

*Coombs, P. H. (ed.) (1980). *Meeting the Basic Needs of the Rural Poor: The Integrated Community-based Approach*, Pergamon Press, New York and Oxford.

Fennell, R. (1979). *The Common Agricultural Policy of the European Community*, Granada Publishing, St Albans.

292

Gardner, B. L., and Richardson, J. W. (eds) (1970). *Consensus and Conflict in United States Agriculture*, Texas A & M University Press (for Texas Agricultural Experiment Station and Agricultural Council of America), College Station, Tx.

Galbraith, J. K. (1970). *The Nature of Mass Poverty*, Penguin, Harmondsworth.

Halcrow, H. G. (1980). *Food Policy for America*, McGraw-Hill, New York.

*Hallett, G. (1981). *The Economics of Agricultural Policy*, 2nd edn, Blackwell Scientific Publications, Oxford.

*Hayami, Y., and Ruttan, V. W. (1971). *Agricultural Development: An International Perspective*, Johns Hopkins Press, Baltimore.

Howe, C. W. (1979). *Natural Resource Economics: Issues, Analysis and Policy*, John Wiley, New York.

*Hunter, G. (1969). *Modernising Peasant Societies*, Oxford University Press, London.

*Hunter, G., Bunting, A. H., and Bottrall, A. (eds) (1978). *Policy and Practice in Rural Development*, Croom Helm, London.

*International Association of Agricultural Economists (three-yearly) *Proceedings of Conferences*, Gower, Farnborough.

*Jackson, W. A. D. (ed.) (1971). *Agrarian Policies and Problems in Communist and Non-communist Countries*, University of Washington Press, Seattle and London.

*Joy, L. (ed.) (1978). *Nutrition Planning: The State of the Art*, IPC Science and Technology Press, Guildford.

Marsh, J. S., and Swanney, P. J. (1980). *Agriculture in the European Community*, Allen and Unwin, London.

Mellor, J. W. (1966). *The Economics of Agricultural Development*, Cornell University Press, Ithaca, NY.

Mellor, J. W. (1984). Agricultural growth: structure and patterns. In *Proceedings of the 18th Conference of the International Association of Agricultural Economists*, Gower, Farnborough, in press.

*Newby, H. (1982). Rural sociology and its relevance to the agricultural economist: a review. *Journal of Agricultural Economics*, **33**, 125–165.

†Overseas Development Institute (1979). Integrated rural development. Briefing paper no. 4, ODI, London.

Paarlberg, D. (1980). *Farm and Food Policy: Issues of the 1980s*, University of Nebraska Press, Lincoln and London.

*Parsons, K. H. (1978). The political economy of agricultural development. Land Tenure Centre Paper No. 116, University of Wisconsin, Madison.

†Ray, S. K., Cummings, R. W., and Herdt, R. W. (1979). *Policy Planning for Agricultural Development*, Tata McGraw-Hill, New Delhi.

*Ritson, C. (1979). *Agricultural Economics: Principles and Policy*, Granada Publishing, St Albans.

Tarrant, J. R. (1980). *Food Policies*, John Wiley, Chichester and New York.

Tweeten, L. G. (1979). *Foundations of Farm Policy*, University of Nebraska Press, Lincoln and London.

Thirlwall, A. P. (1978). *Growth and Development* (with special reference to developing economics), Macmillan, London.

*Schultz, T. W. (ed.) (1978). *Distortions in Agricultural Incentives*, Indiana University Press, Bloomington and London.

Seers, D. (1981). The meaning of development. In *Extension Education and Rural Development*, Vol. 1 (B. R. Crouch and S. Chamala, eds), John Wiley, Chichester, Section 1.

*Sinden, J. A., and Worrell, A. C. (1979). *Unpriced Values: Decisions without Prices*, John Wiley, New York and Chichester.

*Whitby, M. C., and Willis, K. G. (1978). *Rural Resource Development: An Economic Approach*, 2nd edn, Methuen, London.

QUESTIONS AND EXERCISES

1. During the last 20 years, in the nation you know best, what have been the three most important socioeconomic changes affecting agriculture? To what biological and work changes were these socioeconomic changes interrelated?
2. In your own nation, do you judge the flows as indicated in Figure 21.1 to be fully satisfactory? If not, list and briefly explain what improvements you would make.
3. In the area you know best, what differences in value patterns cause the most obvious conflicts (a) between social groups about agricultural policy; (b) between individuals in agricultural production and marketing operations.
4. Give three examples of how lack of 'truth' can lead to faulty economic decisions. Indicate for each example what can be done to correct the position.
5. List ten attributes of 'goodness' that are valued in the society you know best, and that affect agriculture.

Chapter 22

Organization

22.1 INTRODUCTION

The major purposes of organization are to improve the quality of decisions and to speed decision making, the acceptance of decisions, and their effectiveness. Thus responsibility for the innumerable decisions that farm firms make is, in almost all nations, divided out amongst many thousands of firms. The decisions in marketing both products and factors are also divided out in complex structures (see Section 14.3.2 for that of wheat in Canada). Decisions in research operations have to be taken mainly by individuals or small teams. The responsibilities of national governments require decisions about policies to be concentrated in national capitals, but decisions about execution to be dispersed to local centres. *Within* even small firms, and still more within large bodies such as governments and their bureaucracies, organization is essential if goals are to be well related to values, and actions are to achieve goals.

The socioeconomic importance of these major purposes is increasing because the problems of management in firms and governments are made more intricate and often more urgent by changes in technology, greater asset requirements, changing economies of scale, unfavourable price changes, more elaborate infrastructures, advancing education, population pressures, tensions between social groups, social and national security issues, and other concerns.

Yet the importance of understanding the basic purposes and principles of organization is commonly not appreciated in debates of 'development' and 'policy', nor in most college and university courses.

22.2 PURPOSES IN RELATION TO AGRICULTURE

The principal purposes can be summarized as follows:

(a)   The major basic purposes of improving decisions and their acceptance and use require that there are divided out in *appropriate structures*: (i) *responsibilities* for decisions; (ii) the *power and authority* to make them and most likely to secure their acceptance and use. Each group and individual in the structures should have the right role and the right status. We can study the structure of an individual firm; or the overall

294

structure for agricultural production and marketing in a nation's socioeconomic system; or parts of this such as the structure for the marketing of a particular set of products.

(b)   The most appropriate *flow of information* into, within, and out from a structure should be facilitated by it.

(c)   A structure should help to foster the most useful *motivations* of all individuals and groups in it.

(d)   Organization should help facilitate *access* to markets or other outlets for the products of structures and likewise access to factors (including loans). Mobility of persons and resources within and between structures should, on basic economic principles, be so good as to ensure that the marginal products of management, labour, and other factors could not be increased by any further movement. Structures in government bureaucracies, and many colleges and other non-profit-making bodies have *products* (services) and *outlets* (requests for them) even when no market prices can be quoted.

(e)   Organization should help therefore to ensure that *economies of scale and of specialization* are realized, but with due regard to (i) all the alternative ways of organizing to use scarce, 'lumpy' factors; and (ii) transport and communication costs.

(f)   Organization should help to *reduce income fluctuations and strengthen* the abilities of firms to overcome difficulties due to changes in markets and prices.

(g)   Organization should help to ensure future success through promoting (i) education (including training in skills for all manual workers and managers); (ii) appropriate investments and reserves.

22.3   DESIGNING ORGANIZATIONAL STRUCTURES

22.3.1   Procedures

Basic purposes and goals   As for all other designing, the basic purpose of the body to be designed should be defined. If flexible pricing in markets is used, governments and bureaucracies have fewer problems. But their purposes are much extended where central planners fix prices, or intervene greatly to affect prices. Where competition between firms is markedly imperfect, the purposes of firms often become more complex. Non-profit-making bodies have incomplete guidance from prices. We have noted the wide range of purposes that individual farm firms can have (Table 12.1).

In general, the basic purposes of all structures should accord with the values, goals, folk ways, morals, and laws of the societies in which they operate. Where these are changing fast, or being changed, some structures cannot be in complete accord (e.g. the democratic *panchayats* of India). The causes of discords should therefore have close study along with basic purposes.

Commonly old organizations have goals that are better defined than basic purposes. Basic purposes need updating and therefore goals should be reviewed (Sections 12.1 and 15.2).

Activity analysis   If goals are to be reached, the activities that must be well undertaken are almost always more numerous, their types more varied, and their best timings more precise, than is at first understood. And old structures can need modification, because the activities they facilitated are no longer adequate. A comprehensive listing of activities is therefore desirable. For example, for farm firms the list could range from day to day work on enlarged dairy herds in winter, or machinery maintenance, or seasonal operations on new crops, to whole farm planning, or the making of income tax returns.

Decision analysis   For all activities, the types of decision required can be listed, with indications of the kinds of information required for them, and the kinds of consequence they may have within the structure and in relations with other structures. (For example, some decisions of a division within a government department will affect the responsibilities of other divisions, and the relations of the department with other departments and even of the government to the whole social structure. And the consequences of mistakes can arise over many years.)

Relations analysis   For all decisions, the types of information required should be listed, so that the structure design will facilitate all the appropriate flows: into; up and down, and horizontally within; and out – purpose (b). All this should be with due regard to the speed required and to consequences of the delays or inadequacies. Relations analysis should also indicate opportunities to foster useful motivations, mobilities, scale economies, stabilities, and education – the purposes (c)–(g).

22.3.2   Alternative designs

Farm firms   In millions of farm households, the head of the family is legally the farm firm. But for most purposes, the firm is a partnership of husband and wife. Some responsibilities are divided between them and some are met by either, according to how needs arise. The total labour force is small. But the detailed activities are numerous and very varied in kind and timing. So there is little specialization, much flexible changing of duties of individuals, and much team work. But also individuals often carry heavy responsibilities when working on their own (e.g. in dairy herds, sowing grain, using agrochemicals). And long arduous hours have to be worked in some seasons. After tax and interest payments, the net value added has to be shared out largely within the family for consumption or reinvestment. Disagreements and demotivations may occasionally arise from this sharing (Section 16.3.1). But generally, the rewards for manual and managerial efforts and uncertainty bearing are direct. They do not need to be filtered through the awkward wage bargaining and bonus allocating procedures of large firms and governments. The motivations are also great to learn skills, to consider the long term as well as the short, to weather biological, work, or economic difficulties, and to maintain individual freedoms and other values (Table 12.1). These are the reasons why most nations depend on an overall structure for agricultural production that is largely made up of small family firms whose structure commonly secures the advantages of partnership; skilled, flexible team work; and direct motivations of individuals.

But we should also note how important within the overall structure are the many firms and government agencies selling to, buying from, or in other relations with farm firms.

Within the overall structure, the relations between farm firms and other structures can be changed. In the USSR, when the central government concluded that it wanted more direct control of the distribution and use of grain and other outputs, and transfers of assets from agriculture for off-farm development, it secured the establishment of very large collective farms and some state farms, with land and other assets previously used by small family firms. The organizational structure of these very large farms is different from that of family farms. In Israel, socioeconomic values and conditions have resulted in a wide variety of structures for farm production, ranging from completely communal groups to small family firms. Mainland China is trying to evolve an overall structure that on farms will provide adequate motivations while facilitating mechanization and intensification, but still allow government to retain controls, according to their values and goals. As in the USSR and parts of Israel therefore, the structural problems of large bodies have to be faced.

Other examples of changes that determine sizes and structures for farm firms are (a) the successful design of large combine harvesters and other 'lumpy' factors from which the services have to be spread over large crop areas (these large areas must be either on individual large farms, or on numbers of farms that co-operate in groups or buy the services from contracting firms); (b) other specialization so that some functions previously on farms are the responsibility of off-farm units and management problems on farms are therefore simplified (e.g. cheese or butter-making; apple or egg packing; much accounting); (c) changes in price formulation for farm products such that price uncertainties appear reduced; (d) high land prices that, together with lending institutions, restrain purchases by firms with low net worths, so that well established firms become larger; (e) education, with improvements in abilities in management; (f) government policies on land reforms and settlement schemes.

The difficulties of farms that are too small are essentially that the land and other factors in their 'input mix' are insufficient in relation to 'fixed' labour and management, so that incomes are too small. The difficulties of big farm businesses are essentially that 'fixed' management in the 'input mix' is inadequate. They begin to face the problems of all large organizations, and they can be severe because of the nature of agricultural decision making.

Large organizations   Several problems in organization are obviously greater when large structures are compared with small. Within the large, there must be more delegation. Therefore the communication and thorough acceptance of values and goals, and of much information (including prompt 'feedback'), are more important and usually more difficult. Because 'power' has to be shared amongst many people, difficulties are greater in deciding how, and in motivating, training, testing, and rewarding correctly. Even if the specification of responsibilities for all groups and individuals in the structure (job specifications) are set down in detail, who is really responsible for a success or failure often cannot be well judged. Sharing out the net value added often presents crucial difficulties. Maintaining good cohesion becomes more difficult as structures are enlarged. Large organizations are usually held to be 'accountable' to the societies within which they operate, and this raises many problems. Large profit-making firms are accountable to large bodies of equity holders. Government departments and their political heads are accountable to governments and, perhaps through parliaments, to all the people.

The designs of large organizations that will best provide help in reducing these problems and best serve the purposes (a)–(g) are based on results of activity, decision, and relations analysis, and amended and evolved in the light of experiences. Each is therefore in some ways virtually unique. Some general classifications are, however, useful.

In large firms that seek profits measured in money terms, the delegation of major responsibilities may be (a) by *functions* or (b) by *federal decentralization*. Thus with (a), the divisions in a food processing firm, could be (i) finance and accounting; (ii) personnel; (iii) engineering and plant; (iv) raw material procurement; (v) processing operations, including quality control and packaging; (vi) sales and distribution. All products would be the responsibilities of all these divisions. Profits could be assigned only to the firm as a whole. Co-ordination between divisions would depend largely on the Managing Director and group of divisional heads around him and on working groups or 'commando teams' dependent on support from the various divisions. Alternatively with (b), responsibility for profit-making could be decentralized. Separate plants using different raw materials and located in different areas would each have a plant manager and he would be responsible for the contribution of the plant to the firms' total profits. Only some services to the plants would be provided from the federal centre (e.g. guidance on the firm's *general* goals and ethical values, procurement of loans, consultancy services in engineering and microbiology, *some* staff training and career guidance).

In large organizations that are not profit-seeking, the major delegations may be partly by functions and partly by what can be regarded as federal decentralization, even though no profit measures nor accepted measures of net outputs are available. Thus in Scottish universities, financed largely by central government, the major delegation is functional. The Courts are responsible for finance, buildings and land, and the legal aspects of staff employment. The Senates are responsible for teaching and research, but this responsibility is largely decentralized to academic departments according to 'subjects'. These departments compete, even amongst themselves, for students and academic status, but they receive some services from the federal centres (the Courts and Senates). As in almost all other university structures, serious difficulties arise in attempts to deal with multidisciplinary research and teaching. The functions in this and many other interdepartmental and Court–Senate problems have to be delegated to committees. Accountabilities are through various types of communication to the general public, to the body of alumni, and to the government through a federal centre for all the UK universities – the University Grants Committee. This has many advisory committees.

Government   Organization for government is especially difficult and important. Decisions range from choices between alternative designs for minor items of infrastructure (e.g. village toilets) to major policy choices (e.g. in macroeconomic management; providing social and national security). Regional and local differences are often substantial. The attitudes of 'people' are neither constant nor easy to assess well in quantitative terms. Because few if any profit measures are available, and many bureaucracies are large, systems of rewarding and motivating government staffs are often elaborate, but seldom fully satisfactory. Ambitions tend to lead to over-expansion of staff numbers, but unwillingness to be responsible for difficult innovations.

Functions in government are therefore delegated in many interesting ways. These are related to socioeconomic conditions, but with wide differences in success. Major differences in organization affect (i) how central governments are elected or otherwise established, and therefore arrangements for accountability; (ii) the acknowledged responsibilities of political parties; (iii) the sizes, structures, selection, training, reward, and promotion arrangements of the government staffs; (iv) decentralization of responsibilities from centre to provincial and local governments and how these are established and the responsibilities of their staffs; (v) arrangements for finance of government services; (vi) attitudes and practices as affecting information flows within the system to 'government circles' and to the general public; (vii) arrangements for feedback from outside the structure, including the results of independent objective analyses.

Hierarchical and more diffuse structures   In both profit-seeking and other structures, another type of distinction is between (a) hierarchical (pyramidal) structures that can be authoritarian and strictly disciplined and (b) more diffuse structures where 'power' is less concentrated, and smaller nucleii of 'power' may change more readily according to changes in problems. Army organization is commonly given as an example of (a). In armies, lines of communication must be well understood, the division of responsibilities for quick decisions at each level must be clear cut, and discipline throughout must be wholly reliable. Traditional firms that seem to have few 'new' decisions to make are also regarded as hierarchical. More diffuse structures may be found in firms in industries where the pace of research and development for new products is rapid (e.g. for agrochemicals). Actual structures may not be so rigidly classified, and much depends on how structures are used in practice. (For example, in traditional firms, working parties may be set up for particular 'new' types of activity.) In diffuse structures, 'authorities' can become established who tend to obstruct effective mobility of people, ideas, information, and funds.

22.4 SOME TESTS OF STRUCTURES

Structures should be judged so far as possible by their results. Do they accord with the original values and goals, or with accepted and appropriate modifications of these? If there can be profit measures, how have they changed and how do they compare with those for other bodies? Could they probably be improved by better forward planning with, for example, changes in scale, specialization, use of outside services, vertical integration? How well have the purposes in (a)–(g) in Section 22.2 above been achieved?

Related questions that can be useful are often: Are there more than the essential number of levels of management, so that the frictions and obstacles to information flows are excessive? Have excessive management overheads been incurred? Is there excessive insistence that information and requests should flow through only 'proper' channels? At all levels, do managers tend to *enforce* or to *educate*? Are the age distributions of staff inappropriate in relation to future staff requirements and opportunities to recruit? Are promotions ill-judged? Are the external relations of the structure inappropriate?

FURTHER READING

†Arnon, I. (1968). *Organization and Administration of Agricultural Research*, Elsevier, Amsterdam.
*Child, J. (1977). *Organization: A Guide to Problems and Practice*, Harper and Row, London.
Drucker, P. F. (1968). *The Practice of Management*, Pan Books, London.
Food and Agriculture Organization (1979). *Improving the Organization and Administration of Agricultural Development*, Organization and Administration Report No. 1, Food and Agriculture Organization, Rome.
*Hill, F. F. (1964). Institutional development at home and abroad. *Journal of Farm Economics*, **46**, 1087–1094.
*Leonard, D. K. (1977). *Reaching the Peasant Farmer: Organization Theory and Practice in Kenya*, University of Chicago Press, Chicago.
*McLoughlin, P. (1971). *The Farmer, the Politician and the Bureaucrat: Local Government and Agricultural Development in Independent Africa*, Peter McLoughlin Associates, Fredericton, New Brunswick.
Mosher, A. T. (1971). *To Create a Modern Agriculture: Organization and Planning*, Agricultural Development Council, New York.
*Ratman, N. V., and Rao, B. B. (1979). *Management of Social Development in the Rural Area*, Indian Institute of Management, Bangalore.
Schumacher, E. F. (1973). *Small is Beautiful*, Blond and Briggs, London.

QUESTIONS AND EXERCISES

1. Select an organization employing more than 20 people. Describe (a) basic purposes and goals; (b) the activities that should therefore be well undertaken; (c) the distribution of responsibilities of different types within the organization.
2. Contrast two organizations with different structures. State their basic purposes and goals. Explain differences in activities, and in the information flows required for decisions.
3. For the rural area you know best, list the farm-economic and socioeconomic advantages and disadvantages of farm firms with between one and four men, as against those with many more.
4. Apply tests of structure to the organization responsible for the lowest tier of government in your nation. Include both policy formulation and executive responsibilities.

Chapter 23

Some successes and failures

23.1 INTRODUCTION

We noted in Chapter 1 some important statistics of increases in crop and animal production. We classified the ways in which these were secured, and the obstacles to greater changes. And, in later chapters, we noted other examples of symptoms of success. We can now use our overall understandings to examine briefly other examples of particular developments – some judged successful and some unsuccessful. These can serve to show further, in relation to policy issues, the usefulness of our 'systematic' approach.

But when we study past changes in this way, we should beware of pitfalls. We noted, particularly in Chapters 12 and 21, that the complex value patterns of individual farm firms, of societies, and of groups within societies, can differ and can change over time. So some of the criteria of success differ and change. Judgements of success or failure can depend on who is judging, and when. Moreover, at any particular one time and place, many uncertainties have to be faced in formulating goals and plans to achieve them. At some future time of judgement, tracking back so as to understand fully these uncertainties is especially difficult. Also the achieving of one goal often causes new problems and the need for new goals. Success may look like failure. Failures to achieve intended goals can result indirectly in other gains. Failures can be turned into successes. For this short chapter therefore, telling examples have been selected that can be introduced briefly because they are clear enough about (i) goals; (ii) alterations to goals; (iii) whether goals were consistent with basic values and resources; (iv) whether goals were achieved or not; (v) what new or continuing problems resulted and how weighty they seemed.

By selecting such examples we should, of course, be fully aware that the natural and human truths about even them are in many ways far more detailed than short notes on key points can convey. Even more important, we should bear in mind the many reasons why often now we should judge not only detailed questions, but the general core questions of Chapter 1, and on an international scale.

303

23.2 BIOLOGICAL

23.2.1 Successes

NPK as fertilizers   Although field observations had led to the accumulation over centuries of much useful knowledge about soil fertility, progress in finding the 'principle of vegetation', was extremely slow until Jean Boussingault began experiments in Alsace about 1834. Even after Justus von Liebig's book *Chemistry in its Application to Agriculture and Physiology* was published in 1840 and John Lawes patented his process for making rock phosphates soluble, there were 60 years of controversy over the part played by nitrogen. The total research resources used were minimal. By modern standards, therefore, scientific progress was slow. And even today, while major principles are widely known, actual responses to NPK can be reliably predicted within narrow limits on only a small proportion of sites. In very few localities have the whole nitrogen cycle and losses to water courses and their consequences been well quantified. So biological sub-system studies are still far from complete. Moreover, from engineering, farm-economic, and socioeconomic standpoints, developments are still going on, and problems still arising that stem from the basic discoveries of the 1840–1900 period (Sections 2.3, 8.2, 8.3, and 12.2).

But the NPK story may be judged to be one of substantial success. From a socioeconomic standpoint, the goal of discovering the chemical and plant physiological 'principles' was so very important, indeed overdue. And these principles were secured and used, although the general goal had to be detailed more and more as biological complexities were disclosed. Very useful early detailing was by Lawes, who married chemistry to practical farming.

The development of artificial insemination   The goal here was to make possible more rapid genetic improvement of farm animals. It was appropriate in relation to growing understandings of animal genetics, and to the natural and organizational limitations on wider use of rigorously selected sires. The goal was rightly narrowed to give priority to dairy cattle and it was then successfully achieved in many countries. Detailed biological problems arose and also many organizational, farm-economic, and market problems, some of which are not yet fully solved. But useful success can rightly be claimed (Section 5.3.3).

23.2.2 Partial successes

Some biological achievements give rise to such new and continuing problems that we can regard them as only partial successes. A general example is *chemical control of insect pests and fungal diseases*. Many battles have been won biologically, and, with related engineering, farm-economic, and marketing developments, consumers have benefited. But the war still goes on because genetic resistances to particular chemicals increase, natural controls are destroyed or weakened, and chemical controls become increasingly costly. Even genetic resistances in some crop plants to particular insects or diseases can prove inadequate.

Similar partial successes and continuing challenges may be recognized in work on some animal health problems (e.g. mastitis in dairy cows; calf scours).

23.2.3 Failures

Various attempts have been made during the last 50 years to find new ways of *conserving green grass* without losing as much of its nutritive value as is lost in making hay or silage. Drying with the use of energy from fossil fuels has proved too costly, except when the dried product is for special nutritional purposes. Squeezing to remove grass juices and treating the juice and pulp fractions in various ways have been biologically satisfactory, but too slow and costly. The engineering design problems in achieving quicker and cheaper fractionation have not been solved. The older processes of hay and silage making can, despite their biological losses, still out-compete any new process.

In animal production, a goal that has long seemed economically attractive has been to secure on a commercial scale the *reliable production of twin calves* from each cow pregnancy. The high costs of cow production and maintenance would be spread over many more calves. But the natural biological controls of conception and gestation in cattle cannot yet be managed artificially at sufficiently low cost and with sufficient reliability. Man can still not alter genetic controls of reproduction, derived from the evolution of cattle as a species, as readily as he can alter nutritional and other environmental conditions.

23.3   ENGINEERING

23.3.1   Successes

The choice of goals in engineering tends to be quite closely related to the conditions and production systems on particular sites. Thus the development of suitable *tractor–implement linkages* was a goal that Harry Ferguson chose after much field experience of early tractors in Ireland. He had in mind the need for quick and easy control of implements, easy turning, and wheel grip so that lower powered tractors would be adequate. The three-point hydraulic controlled linkage that he designed was recognized as a piece of clever engineering and, from a farm-economic standpoint in western Europe and North America, it made substitution of tractors for horses and some manual labour more reliable and profitable.

Again of course, there were new and continuing problems (e.g. in retraining and training of labour; displacement of labour off farms; much increased dependence on fossil fuels). And in countries with low labour costs and much unemployment and underemployment, tractorization raises serious socioeconomic issues (Sections 16.2.2 and 16.3.2). But Harry Ferguson's goal was well chosen, and we may judge that he was successful.

Low volume spraying of insecticides and fungicides became possible because of imaginative co-operation between biologists and engineers in the scientific design of sprayers. The goal had widespread biological and economic importance. The physical characteristics of materials to be sprayed, and site and plant conditions, were sufficiently specified. Prototypes were well tested in physical terms, and related biological studies then proceeded. Success can, however, only be claimed as regards some of the battles in the continuing war against biotic factors (see Section 23.2.2 above).

23.3.2 Failures

Clear-cut telling examples of these are fewer, although in the evolution of all types of modern equipment faults have been inevitable: corrections and improvements in succeeding models, common. But one type of failure concerns tasks on small farms in the Tropics. *Rice transplanting* is a widespread example. Its design problems are still not solved despite the arduous nature and the disease risks of present methods, despite the urgency of transplanting (especially in relation to some modern cropping systems), and despite substantial attempts and some progress in Japan. Engineering success would undoubtedly lead to needs for farm-economic adjustments and to socioeconomic difficulties, due to rural underemployment and income distribution changes. But in view of the opportunities that success would open up to increase rice production and to ease work, we should probably not judge the challenge of these difficulties to be a valid reason for accepting design failure.

In richer countries, two types of design problem tend now to attract too few resources so that both fail to be solved. The first *concerns smaller farms* and is in some ways similar to the intermediate technology design problems of countries where labour is cheap and capital is dear and scarce. For example, numerous farms in western Europe have labour forces of little more than one man, or are part-time farms, but their tasks in crop and animal production are many and time-consuming. Too few attempts are made to simplify work and to redesign building layouts, and to design cheap equipment of suitable capacity for various tasks. The reason is the assumption that big 'lumpy' items of equipment are inevitable and farm sizes should be increased to match them. But from a socioeconomic standpoint this assumption is not everywhere and always wholly valid. Certainly so long as many farm families value highly the various social and other benefits from small farms, and so long as societies vote to subsidize these farms, the design problems for them should receive more attention.

The second type of engineering failure in richer economies is also due to judgements of net returns on scarce design resources, and to engineering difficulties. It concerns *tasks such as harvesting raspberries and other soft fruit* that are intricate and closely related to long established pruning and other production practices; very important in commercial production of the particular crop; but not very important in relation to agricultural machinery markets as a whole. Success must depend on clever and low cost design, but also on confidence in

306

future expansion of sales of a new design. And this in turn probably depends on willingness radically to alter production practices, as well as on design skills related to alterations that are biologically and economically feasible. So far, such 'circles' have remained unbroken for many special crops, despite some subsidies from governments.

23.4   FARM-ECONOMIC

23.4.1   Successes

In selecting telling examples here, we recognize even more that real success in any one sub-system of agriculture depends on understanding possibilities in other sub-systems and on support from markets and other components of the socioeconomic system. Thus the successful *adjustments* made by thousands of small farm firms in *Denmark and the Netherlands* to changes in international markets for grain and animal products after 1870 have long been used as examples of rational decision making on farms. But government policies were virtually essential on international trade, education and research, land tenure, finance, and law as affecting the working of co-operative groups of farmers. Similarly the successes in achieving what in short we can call 'modern agriculture' in, for example, our Aberdeen, Norfolk, and Cortland groups all depended not only on farmers' decisions but also on the environments in which they could be made – commercial, legal, financial, educational, research, and general social. This does not detract from the economic successes of farm firms themselves. These are indeed all the more clearly recognized when we consider again the variations within the groups in many measures of production, productivity, net returns, and rates of change (Tables 11.1 and 18.1).

In recent decades, agricultural economists have increasingly recognized that many *traditional farm-economic plans in poorer countries* are clever and quite complex solutions within the constraints of particular climates, soils and other natural resources, and particular markets. Sound developments for areas have seemed to depend on improving human and work resources and market opportunities, more than simply on more logic in farm planning. Thus, to give but one example, in our Iloilo group individual farm families have taken their own decisions about cropping and crop production practices, based not only on biological experiences and advice but also on their own land, labour, and capital resources and their own consumption demands (Barlow, 1983). Their past rationality and flexibility provide confidence that worthwhile changes will continue. These human factors are as essential to the 'green revolution' as are biological, engineering, and other factors.

23.4.2   Failures

We can, however, recognize from Parts II, III, and IV that within areas, in both rich and poor countries, *substantial proportions of farmers* have probably failed to achieve their farm-economic goals because these were not well defined, farm planning was inadequate, and human abilities and energies were insufficient. The continuing challenge to education in farm economics is therefore of major importance almost everywhere (Crouch and Chamala, 1981).

Some failures are more serious in that most firms in whole areas are caught in what have been called '*traps*'. Thus the rationality and flexibility of Iloilo rice growers have so far met the challenges of increasing human numbers but have not been able to meet demands for substantially increased incomes per head. Padi rice and other crops, and the limited development of the Philippino socioeconomic system, seem to have trapped these small farmers into a limited pattern of activities and much underemployment. This is a common situation in cultures in which padi rice growing is a major activity. But similar 'traps' have resulted also where upland tropical areas have high human population densities and intensive production of other food crops (e.g. in Kenya, Tanzania, and Uganda in the Kikuyu, Meru, and Kiga tribal areas). For many more rural people to break out of such 'traps', sound employment opportunities off farms are essential, but these are less likely to be adequate so long as farm-economic aspects of national economies are not studied and debated more fully. In many poor countries with populations increasing rapidly, the prospects are that the 'traps' will continue for many decades. Even if 'green revolutions' are successful, the numbers working on farms will keep labour and management productivities and incomes low.

Farm-economic thinking may also be judged inadequate in many of the '*less favoured areas*' *of richer countries*. By developments elsewhere, these areas were out-competed. Radical changes in choices of products and/or sizes of farm should have been made earlier. Thus many of the less fertile Appalachian farms should have been reforested much earlier than they were in the face of competition from the Corn Belt and elsewhere. The depopulation of remote island farms in the west and north of Scotland may well be judged as too long drawn out and painful, and confused by government policies. Biological improvements for these islands are inevitably closely limited by natural endowments of climate and soil. Transport and other infrastructural services are high in cost in relation to potential outputs, and per head of population.

23.5 GENERAL

23.5.1 Settlements

The many problems in settlement of sparsely populated areas provide tests of socioeconomic systems, and tests in the biological, work, and farm-economic sub-systems of agriculture.

The *settlements of the USA, Canada, Australia, and New Zealand* may well be judged to have been successes. The numbers of settlers, increases in crop and animal production, exports, growth in assets on farms, contributions to assets off farms, and tax revenues from agriculture – all these indicated success, and the pace of much development was spectacular. Many causes of success were obvious. The natural endowments made available were great. The crop and animal genotypes that could be drawn on were meantime satisfactory. Weeds, pests, and diseases were not generally overwhelming. Human labour and management energies available were substantial, and complemented by engineering developments affecting the work system on farms. Markets were greatly expanded by transport, engineering, and refrigeration. And governments provided satisfactory infrastructures and services, affecting land tenure and sizes of farms, trade, loans, law and security, and education. Governments also arranged research and extension education to help decision-making on where to settle, what to produce, and how.

Less obvious was how the *chosen organization*, with heavy reliance on family farms, related to the values of the people concerned and motivated them to contribute immeasurable inputs of labour, management, capital, and risk and uncertainty bearing.

Within the overall success, we should of course discern some failures. Early land tenure and labour arrangements, particularly in what are now the south-eastern states of the USA, may be judged unsuccessful in that large plantations and slavery entailed big social problems later. Culture contact problems with American Indians, Australian Aborigines, and Maoris were not fully solved. Law and security were in some areas inadequate. Deflation of general price levels was serious during the period 1872–96. Research and education were for many years inadequate. The choices of products tried in some areas were not suited to their biological and farm-economic conditions. Wind and water erosion were allowed to become serious in some regions. Decades were required to find successful ways of farming the drier regions in North America now largely used for spring-sown wheat. Mistakes were made in stocking the drier lands of Australia. But despite all this, and more, success far outweighed failure.

This may also be our judgement of settlement policy and procedures in the *irrigated areas of the Punjab* (now partly in India) *and the Gezira* (in the Sudan). Land uses and land-use rights of existing inhabitants, the layout requirements of irrigation engineers, and the introduction and establishment of sufficiently profitable production activities all required much socioeconomic information, and much biological, work, and farm-economic information. Substantial research had to be done particularly in the Gezira. Biotic factors were a serious menace. Labour forces had little formal education and were liable to be seasonally too small. Physical conditions for labour were stressful. Processing and marketing facilities for new crops were inadequate.

The organizational structures for all this required to be carefully designed. In the Punjab, heavy reliance was placed on (a) hundreds of thousands of landowners and tenants, (b) the government departments, one of which had overall control of the irrigation water, and (c) many private processing and trading firms. Some farmers' co-operative societies were promoted. In the Gezira, a special corporation (Sudan Plantations Syndicate) was established by government with responsibilities for (i) central services (e.g. irrigation, control of agronomic practices, seed supply, biotic controls, processing, and marketing) and (ii) supervision of land tenure arrangements on which the crucially important policies and initial resettlements were a government responsibility. Government was also responsible for basic policy on labour migration and health, and for much costly biological research and most education. But the special corporation undertook much training, engineering development, and many field trials. Again there was heavy reliance on thousands of small farmers. Risks and uncertainties were borne by the government, the syndicate, and farmers because their returns were, respectively, 40, 20, and 40 per cent shares of the cotton crop, which varied in yield and price.

Some failures were experienced. In the Punjab some drainage was inadequate. Some of the water used for irrigation was unsuitable, and the structures of some soils were allowed to deteriorate. Too little reliance was placed on tube wells. Drainage was not everywhere adequate. Control of wheat diseases was unsatisfactory. Too little research was financed. In the Gezira, the battles against pests and diseases were not always won. The crop rotation insisted on by the corporation became less suitable from farm-economic and socioeconomic standpoints. The original overall division of responsibilities had to be modified to meet changing value patterns. But both the Punjab and the Gezira settlements contributed, and continue to contribute, greatly to human welfare and, overall, can be judged to be successes.

Two settlement schemes in Africa may be judged failures but they have been well described (Baldwin, 1957; Wilde, 1967). In 1931, a plan was approved to create 'an island of prosperity' from irrigation of nearly one million hectares in an almost uninhabited part of the *Middle Niger Basin* in the French Sudan, now Mali. Cotton was to be the main cash crop, and rice the main subsistence crop. Despite massive investment and aid funds from outside the region, less than one-tenth of the intended area was irrigated and net rewards to settlers were too small to attract and maintain sufficient numbers. The reasons were many. From a biological standpoint, water control in the very flat basin was insufficient due to underestimation of rainfall, and engineering difficulties and shortcomings. Weed infestations were allowed to become serious. Crops could not be sown early enough. Research was inadequate. From a work standpoint, labour inputs were too low and mechanization inappropriate as well as costly. Machine maintenance and repair became insufficient. The farm economics were not attractive because small farmers had better opportunities elsewhere. The attempts of the overall controlling body – L'Office du Niger – itself to farm on a big scale using paid labour were generally unsuccessful. This labour was poorly motivated, and the data for sound farm management decisions were inadequate. But 'overhead' costs for management were high. From a socioeconomic standpoint, other obstacles were remoteness from markets, low population densities, the variety of outside sources of loans and aid given for political reasons, and the wide range of functions undertaken by L'Office du Niger.

Our second example of failure in settlement was in another almost uninhabited area of West Africa. In 1947–8, plans were prepared to settle an area of 13 000 hectares of open savannah woodland near 10°N 6°E. Families from densely populated areas of northern Nigeria were to be settled as tenants of small farms and these so laid out that central mechanical and other production services could be provided, as in the Gezira scheme, by a special company, the *Niger Agricultural Project Ltd.* This was owned partly by the Nigerian government and partly by the UK government. The main cash crops were groundnuts and other oil seeds; the main subsistence crop, sorghum. The scheme failed within a few years after only 4000 ha of 'bush' had been cleared. The biological information that was essential was not secured in good time. The work sub-system was inadequate because (i) settlers were not drawn from densely populated rural areas, but, for political reasons, from within the local emirates; (ii) clearing of 'bush' vegetation for mechanized tillage operations was a big and costly task; (iii) for clearing and several other tasks, machinery designs were not suited; (iv) seasonal labour needs for the farm plans could not be met; (v) strict controls of settlers' labour and farm plans were expected, but the company's farm plans were uneconomic on any reasonable assumptions about crop yields and values, fallow requirements to maintain soil fertility, and the 'fixed' costs of crop production in the settlement. Socioeconomic difficulties resulted from (i) 'model' village layouts that were inconsistent with the settlers' social structures and customs; (ii) share tenancies with the Company requiring two-thirds of the crop, so that settlers had insufficient incentives except to cheat.

Failures similar to those in these two settlement scheme examples have continued (Heyer, 1981). Deeper comparisons of them with examples of success are therefore useful. The general conclusions from such comparisons are that (i) however desirable development schemes may seem from a socioeconomic standpoint, they should be based on adequate farm-economic plans; (ii) these require much information about biological, work, farm-economic, and socioeconomic matters; (iii) unless enough of this information is secured by enquiries and small scale trials before wider development is started, serious mistakes will be made in all the sub-systems of agriculture and in the socioeconomic system; (iv) these mistakes may well be too costly and too slowly corrected where too much reliance is placed on particular central corporations or bureaucracies and too little on the direct motivations of thousands of pioneering families; (v) for full success, such families require adequate land tenure arrangements, markets for products and factors (including loans), and other infrastructure, including education and continuing research relevant to their local problems. Perhaps the most frequent mistakes are those that result in plans and controls that are not in accord with the value patterns of farm families (Section 12.1) and their alternative opportunities (Sections 16.2 and 16.3).

23.5.2 'Permanent' crops

Similar conclusions can be drawn from the long history of experiences with tree and bush crops. Much depended on close attention to the particular climatic requirements of the species. So the discovery of the best locations was crucially important. But other biological relationships affecting potential production had also to be learnt before good production practices could be established. There were many biological failures (e.g. in attempts to establish in Assam tea cultivars from China; in early coffee growing in Ceylon (Sri Lanka) and parts of Brazil; in apple orchards on poorly drained soils in New York State). But because the pioneering work was almost entirely by numerous small firms (with some research and other assistance from governments) the whole evolutionary process proceeded to biological results that were generally satisfactory. Labour requirements were high both in establishing plantations and annually. Capital requirements were also high. Special processing was required (e.g. for rubber latex, tea, oil palm fruit). So crucial decisions were made about labour, finance, and organization with significant results in the work and farm-economic sub-systems and socioeconomic system.

Thus in *Malaya* a structure for *rubber production* evolved with smallholder families, plantation companies hiring management and labour and financed by equity shareholders, expert visiting agents supervising the managers, and research and development services (including genetic improvement and latex processing) provided partly under government auspices (with finance from taxes on rubber) and partly by the bigger plantation firms. Such a structure was well supported by land tenure, marketing, labour, and other infrastructural arrangements. It contributed greatly to world supplies of rubber and to making Malaysia and Singapore among the richer nations of Asia.

But the success was not complete. The pace of early planting was so rapid that, in the 1920s, rubber production was restricted to raise prices, so that increased planting in other countries was promoted. After 1954, when cultivars were available with two to three times the yield of older cultivars, replanting on smallholdings was retarded by the complex patterns of plot tenancies, some joint ownerships of trees, and the biological losses in trying to establish young trees in small plots due to shading by old trees on surrounding plots. Labour and land tenure policies were considered by the 1950s to require revision, because Malay people were not 'progressing' as fast as those with Chinese origins. For ethnic reasons too, special new plantation settlements were felt necessary, even although such big government-supported and bureaucratically managed settlements could lead to the dangers evident from our African failures (Barlow, 1978).

The great build-up of knowledge and organizational structure for the Malaysian rubber industry served to limit any failures. But we can usefully note that such failures occurred in *Mexico, Brazil, and the Philippines*, when in the 1920s bureaucratic schemes were introduced for rapid rubber planting. As in the African settlement examples, the causes were partly biological (e.g. fungal diseases), partly in securing labour, partly farm-economic, and partly inadequate socioeconomic environments.

Smaller plantation firms, however, also failed even where socioeconomic environments was generally in their favour. For example, the planting of *coffee in Ceylon* proceeded during the eighteenth century at a pace that was too rapid in relation to the essential build-up of biological knowledge and farm-economic skills. The results were much disease (leaf rust) and many bankruptcies, but also some management experiences and other assets on which the successful Ceylon tea industry could be partly based.

23.5.3 Special problems

In some socioeconomic situations, national goals for agriculture tend to be specified more closely. Three telling examples start with the goals formulated during 1939–41 in the UK, during the late 1940s and early 1950s in China, and in later years in Iran. We can briefly trace actual developments in all three cases up to recent times.

United Kingdom   After the outbreak of World War II in Europe in 1939 imports of food and feedingstuffs to the UK, which had been 23 million tonnes a year, were greatly reduced by torpedoes and bombs and by the need to use more ships for military supplies. It was urgently necessary, therefore, to ensure that agriculture adjusted the total quantum and composition of output so as to make the most appropriate contributions to human nutrition. Food imports could be altered in composition, many foods could be rationed, but a major aim must be to 'save ships'.

By 1945, imports of food and feedingstuffs were only about 10 million tonnes but human nutrition was, from a biological standpoint, better than prior to the war. Agriculture contributed to this success, not because the basic structure of small firms was altered but largely because government (i) determined a sound basic strategy of increasing grain, potato, and vegetable production, maintaining milk production, and reducing the resources for commercial meat and egg production; (ii) guaranteed forward prices for products and controlled prices of factors in support of this strategy; (iii) facilitated increased use of fertilizers, labour, and machinery; (iv) compelled farmers to plough up grassland and expand grain and potato hectarages; (v) supervised closely management on some 5 per cent of farms where it was judged inadequate. Thus the government intervened in various co-ordinated ways, but without causing major changes in organization.

The success can, of course, be judged imperfect. Even more ships could have been 'saved' if the shift to vegetarian diets had been greater. Some pests increased (e.g. potato eelworms). The productivity of labour and machinery was not, by peace-time criteria, increased. The system of subsidies that made possible differences between import, farm, and retail price levels became established, although it might not be well suited to post-war socioeconomic conditions. Transfers of farms to the more able potential tenants were retarded. Evolution of marketing structures was halted. Too little was done to counter misinterpretations of results being used to support post-war plans for large scale mechanized farming by government agencies in Africa (including the notorious Groundnut Scheme in Tanganyika (in 1947–52).

In 1947, when importing food was still restricted by shortages and payment difficulties, war-time food rationing was continued and further expansion of production sought. War-time pricing arrangements were continued. The aim of 'saving ships' gave way to that of 'saving foreign exchange and contributing to the end of rationing'. Output increased rapidly because (i) product prices were guaranteed; (ii) subsidies were substantial; (iii) supplies of biological inputs and equipment increased; (iv) a backlog of useful biological work and farm-economic knowledge had accumulated during the prewar depression years but could not be fully used in war-time; (v) the consultancy and extension education services were active and free to farm firms; (vi) the net worths of these firms had increased and interest rates on loans were low. By 1954 food rationing ended. By 1958 it became increasingly necessary to limit subsidies, and so market prices began to return to their prewar importance in farm-economic decision making. But average costs of production had become high. Biological and work skills increased further and the government sustained enough confidence in their economic support of farm incomes so that outputs continued to rise.

Now the UK farm economy is within the European Economic Community (EEC) and farm product prices again depend increasingly on market supply and demand forces, but still quite largely on bureaucratic and political decisions about interventions. These decisions cause much 'confusion and conflict' because they relate to goals that are more varied and controversial than the UK goals of 'ship-saving' in war-time, or of saving foreign exchange in the 1950s.

China In the late 1940s China faced many problems that had accumulated during the nineteenth century, more recent decades of unstable government and inadequate infrastructure, a decade of war with Japan, and wild inflation. The goals formulated therefore related to many aspects of life. Again those for agriculture could be well understood only in the general socioeconomic context. This raised strong popular demands for (a) correction of past inadequacies in government; (b) higher material incomes; (c) lower risks of natural disasters due to flood or drought; (d) more equity in wealth and income distributions and social status. To cater for all these demands, industrialization and use of modern technologies were widely recognized as essential. Enormous increases in capital would therefore be necessary, but there was strong opposition to finance and management from overseas. Human population pressures were already high and rising rapidly.

In these circumstances the main goals of government for agriculture were (i) to increase output rapidly; (ii) to reduce the threat of floods and droughts; (iii) to secure products cheaply so as to permit the build-up of assets, particularly for heavy industry but also for agriculture; (iv) to release some labour for non-farm work but, for some years, to employ many more workers in agriculture because of the pace of population increase and the desire to avoid aggravating urban problems; (v) to provide outlets for industry's production of farm machinery and other manufactures. All these seemed, meantime, more important than raising the general level of personal incomes on farms. But great emphasis was placed on the desire (vi) to secure greater equity.

The main means chosen by government to achieve these goals were radical changes in organization and structure. By 1952, farm land ownership and use rights were redistributed and collectivization began. By 1958, this was into very large '*communes*' within which there were 'production brigades', and within these 'production teams'. The prices of products and factors were mainly decided by government in relation to central food procurement and export plans and overall capital accumulation and investment strategy. Government also controlled labour movements and decreed that Provinces should struggle to be largely self-sufficient in food.

The biological achievements were great. The gross output of agriculture almost doubled between 1952 and 1977. The irrigated area was increased from 21 to 48 million hectares. Almost 7 million hectares of sloping ground were terraced. Flood controls were installed to protect wide areas. Inorganic fertilizers and other agrochemicals were increased from 2.5 to 28.7 per cent of gross output. The use of organic fertilizers was probably about doubled. For the main crops, high yielding cultivars became the most widely used. Plant disease controls were greatly improved. The pig population was increased almost 200 per cent; sheep and goats by almost 140 per cent (Tang and Stone, 1980; Zhan Wu, 1981).

The inputs of work were also great. Between 1952 and 1977 the agricultural labour force increased from 169 to 262 million head; the number of draught animals from 78 to 110 million; the total horsepower rating in tractors and other motorized equipment from 0.2 to 66 million h.p.

From a farm-economic standpoint, however, there were unfavourable trends of total inputs in relation to outputs, and of price relationships. Farm family consumption levels were kept low. Also accounting and analytical practices were inadequate for valuation of water control and other long term investments, as well as for many decisions in management and government policy.

From a socioeconomic standpoint the main goals (i)–(vi) were all achieved, but not so fully and well that the problems in meeting popular demands (a)–(d) for the future seemed much reduced. The reasons are difficult to summarize, but the following appeared to be the most significant: (a) the high population pressures and the low levels of productivity in 1947 and even before the Japanese war and the high rate of population increase; (b) the inevitable gradual decline, post-war and post-revolution, in motivations to produce, pay taxes, and invest without immediate improvements in material rewards; (c) the relatively low priorities for agriculture in government economic plans for 1957–62 and 1965–71; (d) political instabilities arising from uncertainties about value patterns and therefore about the effects on motivations to work, manage, save, and invest of (i) material incomes for personal consumption as against (ii) non-material benefits related to personal valuations of national economic and social changes; (iii) insistence on provincial self-sufficiencies in food supplies and control of labour migrations, so that poor and thickly populated areas had to produce more grain rather than livestock and trees, and soil erosion increased; (iv) bureaucratic control of prices so that farm-economic and socioeconomic decisions could not be based sufficiently on marginal real values; (v) continuing infrastructural deficiencies (e.g. for transport; statistics).

In recent years the economic planning of China's government has been modified in various ways to secure greater motivations again and more true economy in resource use. One major change is the build-up of 45 agricultural universities and colleges. Another is heavier reliance on the 'production teams' of 'communes' to make farm planning decisions, and on the non-farm enterprises of 'communes' and 'production brigades' of 'communes' (e.g. in processing, transport, and even mining and machinery part manufacture). The uses of scientific skills and even some finance from overseas are tending to increase.

So China's experiences since 1947 include much success, but many difficulties remain in achieving more fully the basic popular aims. And experience has shown, in China as elsewhere, that for sound development of agriculture decisions on and off farms need to be altered flexibly and objectively as socioeconomic conditions and value systems change, and knowledge and resources in biological, work, and farm-economic sub-systems change.

Iran Problems and experiences since 1950 in Iran are also internationally significant. They emphasize particularly the socioeconomic importance of (i) value systems; (ii) decisions about the pace of development; (iii) organization and management (Baldwin, 1967; Rassul, 1978).

The basic conditions were those of rapid population increase, closely limited water and soil resources, great lack of formal education and research, and low average personal incomes despite natural reserves of oil and gas. Large landowners had much economic and political power. The value systems of the major political groups were largely incompatible because of past culture contacts and mixtures and more recent international contacts and pressures. (For example, the large religious group was opposed to modern consumer-orientated development; participation in money markets; and taxes to finance many government services as against taxes for religious purposes. Government circles wanted 'Westernization'; rapid increases in material incomes; and population control by abortion as well as other means. Various other groups ranged from liberal-democratic to strictly communist. Because oil reserves seemed likely to be depleted within 40 years, economic planners sought rapid development and government funds could be provided from large current revenues from oil, and loans and aid from overseas.)

As affecting agriculture the goals were to increase output rapidly; to maintain self-sufficiency in major foodstuffs and even to increase agricultural exports; to increase reliance on Iranian manufacturers in substitution for imports; to reduce underemployment in rural areas; to secure more equitable distributions of wealth in land and of personal incomes; and to improve the infrastructure, including education and social services. But such goals were not defined in quantitative terms until the Fourth Plan for the five years 1967–72, when stress was placed also on trying to secure better organization and administration for agricultural productivity. The Fifth Plan, for 1973–8, became even more ambitious, partly because revenues from oil were raised by the OPEC cartel's prices. The plan included a 7 per cent annual growth rate for agricultural output, although this was recognized as requiring major changes in attitudes, bureaucratic regulations, and extension and other services. Significantly also this plan included, as a last aim, 'to conserve and revive the ancient culture of Iran'.

A very wide range of 'laws' was used to change agriculture and the overall result was that, in real terms, the contribution of agriculture to the gross domestic product seemed to increase by 80 per cent between 1960 and 1975. But the total demand in the economy was increased by about 379 per cent and, with urbanization and the policy followed on food prices, consumption of foodstuffs increased so greatly that net imports rose from almost nil to 26 per cent of total consumption. Despite high current revenues from oil, serious difficulties in balancing international payments arose, and were regarded as largely due to the relatively slow increase in agricultural production.

The reasons why the increase was not greater despite all the 'laws' in support were many, but could be classified as strategic, tactical, and human. The *strategy* of the plans favoured urbanization and industrialization more than agriculture and improvements in the socioeconomic environment for agriculture. Mistaken concepts of the economies of scale in irrigation and agriculture prevailed. The *tactics* adopted by the many government departments and agencies set up were not well co-ordinated. Information and skills were lacking on many matters from potential irrigation water supplies or output–input relations in crop production to farm machinery maintenance and the farm economics of pest and disease control. Not much research work was problem orientated. Extension workers were regarded as government officials rather than helpful educators. Land levelling for better irrigation, and loans to help improve 'input mixes' were often delayed after tenants were allocated land in 'land reform'. The *human* reasons affected both strategy and tactics at all levels of staffing in the bureaucracies, and the reactions of farm families to them. 'Confusions and conflicts' over value patterns were fundamental and were aggravated by generally low levels of personal incomes, under- and unemployment, and disparities in incomes. And, particularly during the periods of the Fourth and Fifth Plans, the conflicts were sharpened by the pace of increase in some urban incomes, but the relative depression of agricultural incomes by reliance on food imports and subsidies to keep market prices relatively low. The government attempted to reduce internal conflicts as well as external threats by military expenditures. By 1975, these were 31 per cent of total government expenditure.

The outcome was the revolution of 1979 that installed the Ayatollah Khoumeini as chief exponent of the most common value system, which the government had not sufficiently respected when coming to decisions about pace, priorities, organization, and methods in development.

REFERENCES

Baldwin, G. B. (1967). *Planning and Development in Iran*, Johns Hopkins University Press, Baltimore.

Baldwin, K. D. S. (1957). *The Niger Agricultural Project*, Basil Blackwell, Oxford.

Barlow, C. (1978). *The Natural Rubber Industry*, Oxford University Press, Kuala Lumpur.

Crouch, B. R., and Chamala, S. (1981). *Extension Education and Rural Development, Vol. 2, International Experiences in Strategies for Planned Development*, John Wiley, Chichester.

Heyer, J. (1981). 'Rural development programmes and impoverishment; some experiences in tropical Africa'. In *Rural Change: The Challenge for Agricultural Economists* (G. Johnson and A. Maunder, eds), Gower (for the International Association of Agricultural Economists), Farnborough.

Liebig, J. von (1840). *Chemistry in its Application to Agriculture and Physiology*, translated by Lyon Playfair, Taylor and Walton, London.

Rassul, M. Rezai (1978). *Agricultural development and policies in Iran*, MSc thesis, Aberdeen University.

Tang, A. M., and Stone, B. (1980). *Food Production in the Peoples Republic of China*, International Food Policy Research Institute, Washington, DC.

Wilde, J. C. de (1967). *Experiences in Agricultural Development in Tropical Africa*, Vol. 2, Johns Hopkins University Press, Baltimore.

Zhan Wu (1981). 'The development of socialist agriculture in China'. In *Rural Change: The Challenge for Agricultural Economists* (G. Johnson and A. Maunder, eds), Gower (for the International Association of Agricultural Economics), Farnborough.

QUESTIONS AND EXERCISES

1. Select a goal in the agricultural policy of the government of your country 10 years ago. Define the main types of biological, work, farm-economic, and socioeconomic changes that were required to secure this goal. Indicate whether each has been 'fully secured', 'partly secured', or 'not secured'.
2. Briefly describe two examples of apparent or partial success in agricultural development that led to new problems and goals.
3. Briefly describe two examples of altered assessments of 'the national interest' by governments. Indicate any inadequacies in the resulting policy changes.

Index